The Future Computed

Artificial Intelligence and its role in society

By Microsoft

Foreword by Brad Smith and Harry Shum

The Future Computed

Artificial Intelligence and its role in society

By Microsoft

With a foreword by
Brad Smith and Harry Shum

Published by Microsoft Corporation
Redmond, Washington. U.S.A.
2018

First published 2018 by Microsoft Corporation
One Microsoft Way
Redmond, Washington 98052

ISBN 978-1-9802344-3-2

Table of contents

Twenty years ago, we both worked at Microsoft, but on opposite sides of the globe. In 1998, one of us was living and working in China as a founding member of the Microsoft Research Asia lab in Beijing. Five thousand miles away, the other was based at the company's headquarters, just outside of Seattle, leading the international legal and corporate affairs team. While we lived on separate continents and in quite different cultures, we shared a common workplace experience within Microsoft, albeit with differing routines before we arrived at the office.

At that time in the United States, waking to the scent of brewing coffee was a small victory in technology automation. It meant that you had remembered to set the timer on the programmable coffee maker the night before. As you drank that first cup of coffee, you typically watched the morning news on a standard television or turned the pages of the local newspaper to learn what had happened while you slept. For many people a daily diary was your lifeline, reminding you of the coming day's activities: a morning meeting at the office, dial-in numbers and passcodes for conference calls, the address for your afternoon doctor's appointment, and a list of to-dos including programming the VCR to record your favorite show. Before you left for the day, you might have placed a few phone calls (and often left messages on answering machines), including to remind sitters when to pick up children or confirm dinner plans.

Twenty years ago, for most people in China, an LED alarm clock was probably the sole digital device in your bedroom. A bound personal calendar helped you track the day's appointments, addresses, and phone numbers. After sending your kids off to school, you likely caught up on the world's happenings from a radio broadcast while you ate a quick breakfast of soya milk with Youtiao at your neighborhood restaurant. In 1998, commuters in Beijing buried their noses in newspapers and books – not smartphones and laptops – on the crowded trains and buses traveling to and from the city's centers.

But today, while many of our fundamental morning routines remain the same, a lot has also changed as technology has altered how we go about them. Today a morning in Beijing is still different from a morning in Seattle, but not as different as it used to be. Consider for a moment that in both places the smartphone charging on your bedside table is the device that not only wakes you, but serves up headlines and updates you on your friends' social lives. You read all the email that arrived overnight, text your sister to confirm dinner plans, update the calendar invite to your sitter with details for soccer practice, and then check traffic conditions. Today, in 2018, you can order and pay for a double skinny latte or tea from Starbucks and request a ride-share to drive you to work from that same smartphone.

Compared with the world just 20 years ago, we take a lot of things for granted that used to be the stuff of science fiction. Clearly much can change in just two decades.

Twenty years from now, what will your morning look like? At Microsoft, we imagine a world where your personal digital assistant Cortana talks with your calendar while you sleep. She works with your other smart devices at home to rouse you at the end of a sleep cycle when it's easiest to wake and ensures that you have plenty of time to shower, dress, commute and prepare for your first meeting. As you get ready, Cortana reads the latest news, research reports and social media activity based on your current work, interests and tasks, all of which she gleaned from your calendar, meetings, communications, projects and writings. She updates you on the weather, upcoming meetings, the people you will see, and when you should leave home based on traffic projections.

Acting on the request you made a year before, Cortana also knows that it's your sister's birthday and she's ordered flowers (lilies, your sister's favorite) to be delivered later that day. (Cortana also reminds you about this so that you'll know to say, "you're welcome" when your sister thanks you.) Cortana has also booked a reservation for a restaurant that you both like at a time that's convenient for both of your schedules.

In 2038, digital devices will help us do more with one of our most precious commodities: time.

In 20 years, you might take your first meeting from home by slipping on a HoloLens or other device where you'll meet and interact with your colleagues and clients around a virtual boardroom powered by mixed reality. Your presentation and remarks will be translated automatically into each participant's native language, which they will hear through an earpiece or phone. A digital assistant like Cortana will then automatically prepare a summary of the meeting with tasks assigned to the participants and reminders placed on their schedules based on the conversation that took place and the decisions the participants made.

In 2038, a driverless vehicle will take you to your first meeting while you finalize a presentation on the car's digital hub. Cortana will summarize research and data pulled from newly published articles and reports, creating infographics with the new information for you to review and accept. Based on your instructions, she'll automatically reply to routine emails and reroute those that can be handled by others, which she will request with a due date based on the project timeline. In fact, some of this is already happening today, but two decades from now everyone will take these kinds of capabilities for granted.

Increasingly, we imagine that a smart device will monitor your health vitals. When something is amiss, Cortana will schedule an appointment, and she will also track and schedule routine checkups, vaccines and tests. Your digital assistant will book appointments and reserve time on your calendar on days that are most convenient. After work a self-driving car will take you home, where you'll join your doctor for a virtual checkup. Your mobile device will take your blood pressure, analyze your blood and oxygen level, and send the results to your doctor, who will analyze the data during your call. Artificial intelligence will help your doctor analyze your results using more than a terabyte of health data, helping her accurately diagnose and prescribe a customized treatment based on your unique physiological traits. Within a few hours, your medication will arrive at your door by drone, which Cortana will remind you to take. Cortana will also monitor your progress and, if you don't improve, she'll ask your permission to book a follow-up appointment with the doctor.

When it's time to take a break from the automated world of the future, you won't call a travel agent or even book online your own flight or hotel as you do today. You'll simply say, "Hey, Cortana, please plan a two-week holiday." She'll propose a custom itinerary based on the season, your budget, availability and interests. You'll then decide where you want to go and stay.

Looking back, it's fascinating to see how technology has transformed the way we live and work over the span of twenty years. Digital technology powered by the cloud has made us smarter and helped us optimize our time, be more productive and communicate with one another more effectively. And this is just the beginning.

Before long, many mundane and repetitive tasks will be handled automatically by AI, freeing us to devote our time and energy to more productive and creative endeavors. More broadly, AI will enable humans to harness vast amounts of data and make breakthrough advances in areas like healthcare, agriculture, education and transportation. We're already seeing how AI-bolstered computing can help doctors reduce medical mistakes, farmers improve yields, teachers customize instruction and researchers unlock solutions to protect our planet.

But as we've seen over the past 20 years, as digital advances bring us daily benefits they also raise a host of complex questions and broad concerns about how technology will affect society. We have seen this as the internet has come of age and become an essential part of our work and private lives. The impact ranges from debates around the dinner table about how distracting our smartphones have become to public deliberations about cybersecurity, privacy, and even the role social media plays in terrorism. This has given birth not just to new public policies and regulations, but to new fields of law and to new ethical considerations in the field of computer science. And this seems certain to continue as AI evolves and the world focuses on the role it will play in society. As we look to the future, it's important that we maintain an open and questioning mind while we seek to take advantage of the opportunities and address the challenges that this new technology creates.

The development of privacy rules over the past two decades provides a good preview of what we might expect to see more broadly in the coming years for issues relating to AI. In 1998, one would have been hard-pressed to find a full-time "privacy lawyer." This legal discipline was just emerging with the advent of the initial digital privacy laws, perhaps most notably the European Community's Data Protection Directive, adopted in 1995. But the founding of the International Association of Privacy Professionals, or IAPP, the leading professional organization in the field, was still two years away.

Today, the IAPP has over 20,000 members in 83 countries. Its meetings take place in large convention centers filled with thousands of people. There's no shortage of topics for IAPP members to discuss, including questions of corporate responsibility and even ethics when it comes to the collection, use, and protection of consumer information. There's also no lack of work for privacy lawyers now that data protection agencies — the privacy regulators of our age — are operating in over 100 countries. Privacy regulation, a branch of law that barely existed two decades ago, has become one of the defining legal fields of our time.

What will the future bring when it comes to the issues, policies and regulations for artificial intelligence? In computer science, will concerns about the impact of AI mean that the study of ethics will become a requirement for computer programmers and researchers? We believe that's a safe bet. Could we see a Hippocratic Oath for coders like we have for doctors? That could make sense. We'll all need to learn together and with a strong commitment to broad societal responsibility. Ultimately the question is not only what computers can do. It's what computers should do.

Similarly, will the future give birth to a new legal field called "AI law"? Today AI law feels a lot like privacy law did in 1998. Some existing laws already apply to AI, especially tort and privacy law, and we're starting to see a few specific new regulations emerge, such as for driverless cars. But AI law doesn't exist as a distinct field. And we're not yet walking into conferences and meeting people who introduce themselves as "AI lawyers." By 2038, it's safe to assume that the situation will be different. Not only will there be AI lawyers practicing AI law, but these lawyers, and virtually all others, will rely on AI itself to assist them with their practice.

The real question is not whether AI law will emerge, but how it can best come together — and over what timeframe. We don't have all the answers, but we're fortunate to work every day with people who are asking the right questions. As they point out, AI technology needs to continue to develop and mature before rules can be crafted to govern it. A consensus then needs to be reached about societal principles and values to govern AI development and use, followed by best practices to live up to them. Then we're likely to be in a better position for governments to create legal and regulatory rules for everyone to follow.

This will take time — more than a couple of years in all likelihood, but almost certainly less than two decades. Already it's possible to start defining six ethical principles that should guide the development and use of artificial intelligence. These principles should ensure that AI systems are fair, reliable and safe, private and secure, inclusive, transparent, and accountable. The more we build a detailed understanding of these or similar principles — and the more technology developers and users can share best practices to implement them — the better served the world will be as we begin to contemplate societal rules to govern AI.

Today, there are some people who might say that ethical principles and best practices are all that is needed as we move forward. They suggest that technology innovation doesn't really need the help of regulators, legislators and lawyers.

While they make some important points, we believe this view is unrealistic and even misguided. AI will be like every technology that has preceded it. It will confer enormous benefits on society. But inevitably, some people will use it to cause harm. Just as the advent of the postal service led criminals to invent mail fraud and the telegraph was followed by wire fraud, the years since 1998 have seen both the adoption of the internet as a tool for progress and the rise of the internet as a new arena for fraud, practiced in increasingly creative and disturbing ways on a global basis.

We must assume that by 2038, we'll grapple with the issues that arise when criminal enterprises and others use AI in ways that are objectionable and even harmful. And undoubtedly other important questions will need to be addressed regarding societally acceptable uses for AI. It will be impossible to address these issues effectively without a new generation of laws. So, while we can't afford to stifle AI technology by adopting laws before we understand the issues that lie ahead of us, neither can we make the mistake of doing nothing now and waiting for two decades before getting started. We need to strike a balance.

As we consider principles, policies and laws to govern AI, we must also pay attention to AI's impact on workers around the globe. What jobs will AI eliminate? What jobs will it create? If there has been one constant over 250 years of technological change, it has been the ongoing impact of technology on jobs — the creation of new jobs, the elimination of existing jobs, and the evolution of job tasks and content. This too is certain to continue with the adoption of AI.

Will AI create more jobs than it will eliminate? Or will it be the other way around? Economic historians have pointed out that each prior industrial revolution created jobs on a net basis. There are many reasons to think this will also be the case with AI, but the truth is that no one has a crystal ball.

It's difficult to predict detailed employment trends with certainty because the impact of new technology on jobs is often indirect and subject to a wide range of interconnected innovations and events. Consider the automobile. One didn't need to be a soothsayer to predict that the adoption of cars would mean fewer jobs producing horse-drawn carriages and new jobs manufacturing automobile tires. But that was just part of the story.[1]

The transition to cars initially contributed to an agricultural depression that affected the entire American economy in the 1920s and 1930s. Why? Because as the horse population declined rapidly, so did the fortunes of American farmers. In the preceding decade roughly a quarter of agricultural output had been used to feed horses. But fewer horses meant less demand for hay, so farmers shifted to other crops, flooding the market and depressing agricultural prices more broadly. This agricultural depression impacted local banks in rural areas, and then this rippled across the entire financial system.

Other indirect effects had a positive economic impact as the sale of automobiles led to the expansion of industry sectors that, at first glance, appear disconnected from cars. One example was a new industry to provide consumer credit. Henry Ford's invention of the assembly line made cars affordable to a great many families, but cars were still expensive and people needed to borrow money to pay for them. As one historian noted, "installment credit and the automobile were both cause and consequence of each other's success."[2] In short, a new financial services market took flight.

Something similar happened with advertising. As passengers traveled in cars driving 30 miles per hour or more, "a sign had to be grasped instantly or it wouldn't be grasped at all."[3] Among other things, this led to the creation of corporate logos that could be recognized immediately wherever they appeared.

Consider the indirect impact of the automobile on the island of Manhattan alone. The cars driving down Broadway contributed to the creation of new financial jobs on Wall Street and new advertising positions on Madison Avenue. Yet there's little indication that anyone predicted either of these new job categories when cars first appeared on city streets.

One of the lessons for AI and the future is that we'll all need to be alert and agile to the impact of this new technology on jobs. While we can predict generally that new jobs will be created and some existing jobs will disappear, none of us should develop such a strong sense of certainty that we lose the ability to adapt to the surprises that probably await us.

But as we brace ourselves for uncertainty, one thing remains clear. New jobs will require new skills. Indeed, many existing jobs will also require new skills. That is what always happens in the face of technological change.

Consider what we've seen over the past three decades. Today every organization of more than modest size has one or more employees who support its IT, or information technology. Very few of these jobs existed 30 years ago. But it's not just IT staff that needed to acquire IT skills. In the early 1980s, people in offices wrote with a pen on paper, and then secretaries used typewriters to turn that prose into something that was actually legible. By the end of the decade, secretaries learned to use word processing terminals. And then in the 1990s, everyone learned to do their own writing on a PC and the number of secretaries declined. IT training wasn't just reserved for IT professionals.

In a similar way, we're already seeing increasing demand for new digital and other technical skills, with critical shortages appearing in some disciplines. This is expanding beyond coding and computer science to data science and other fields that are growing in importance as we enter the world's Fourth Industrial Revolution. More and more, this isn't just a question of encouraging people to learn new skills, but of finding new ways to help them acquire the skills they will need. Surveys of parents show that they overwhelmingly want their children to have the opportunity to learn to code. And at Microsoft, when we offer our employees new courses on the latest AI advances, demand is always extremely high.

The biggest challenges involve the creation of ways to help people learn new skills, and then rethinking how the labor market operates to enable employers and employees to move in more agile ways to fill new positions. The good news is that many communities and countries have developed new innovations to address this issue, and there are opportunities to learn from these emerging practices. Some are new approaches to longstanding programs, like Switzerland's successful youth apprenticeships. Others are more recent innovations spurred by entities such as LinkedIn and its online tools and services and nonprofit ventures like the Markle Foundation's Skillful initiative in Colorado.

The impact of AI, the cloud and other new technologies won't stop there. A few decades ago, workers in many countries mostly enjoyed traditional employer-employee relationships and worked in offices or manufacturing facilities. Technology has helped upend this model as more workers engage in alternative work arrangements through remote and part-time work, as contractors or through project-based engagements. And most studies suggest that these trends will continue.

For AI and other technologies to benefit people as broadly as possible, we'll need to adapt employment laws and labor policies to address these new realities. Many of our current labor laws were adopted in response to the innovations of the early 20th century. Now, a century later, they're no longer suited to the needs of either workers or employers. For example, employment laws in most countries assume that everyone is either a full-time employee or an independent contractor, making no room for people who work in the new economy for Uber, Lyft or other similar services that are emerging in every field from tech support to caregiving.

Similarly, health insurance and other benefits were designed for full-time employees who remain with a single employer for many years. But they aren't as effective for individuals who work for multiple companies simultaneously or change jobs more frequently. Our social safety net — including the United States' Social Security system — is a product of the first half of the last century. There is an increasingly pressing need to adapt these vital public policies to the world that is changing today.

As we all think about the future, the pace of change can feel more than a little daunting. By looking back to technology in 1998, we can readily appreciate how much change we've lived through already. Looking ahead to 2038, we can begin to anticipate the rapid changes that lie ahead — changes that will create opportunities and challenges for communities and countries around the world.

For us, some key conclusions emerge.

First, the companies and countries that will fare best in the AI era will be those that embrace these changes rapidly and effectively. The reason is straightforward: AI will be useful wherever intelligence is useful, helping us to be more productive in nearly every field of human endeavor and leading to economic growth. Put simply, new jobs and economic growth will accrue to those that embrace the technology, not those that resist it.

Second, while we believe that AI will help improve daily life in many ways and help solve big societal problems, we can't afford to look to this future with uncritical eyes. There will be challenges as well as opportunities. This is why we need to think beyond the technology itself to address the need for strong ethical principles, the evolution of laws, the importance of training for new skills, and even labor market reforms. This must all come together if we're going to make the most of this new technology.

Third, we need to address these issues together with a sense of shared responsibility. In part this is because AI technology won't be created by the tech sector alone. At Microsoft we're working to "democratize AI" in a manner that's similar to the way we "democratized the PC." Just as our work that started in the 1970s enabled organizations across society to create their own custom applications for the PC, the same thing will happen with AI. Our approach to AI is making the fundamental AI building blocks like computer vision, speech, and knowledge recognition available to every individual and organization to build their own AI-based solutions. We believe this is far preferable to having only a few companies control the future of AI. But just as this will spread broadly the opportunity for others to create AI-based systems, it will spread broadly the shared responsibility needed to address AI issues and their implications.

As technology evolves so quickly, those of us who create AI, cloud and other innovations will know more than anyone else how these technologies work. But that doesn't necessarily mean that we will know how best to address the role they should play in society. This requires that people in government, academia, business, civil society, and other interested stakeholders come together to help shape this future. And increasingly we need to do this not just in a single community or country, but on a global basis. Each of us has a responsibility to participate — and an important role to play.

All of this leads us to what may be one of the most important conclusions of all. We're reminded of something that Steve Jobs famously talked about repeatedly: he always sought to work at the intersection of engineering and the liberal arts.

One of us grew up learning computer science and the other started in the liberal arts. Having worked together for many years at Microsoft, it's clear to both of us that it will be even more important to connect these fields in the future.

At one level, AI will require that even more people specialize in digital skills and data science. But skilling-up for an AI-powered world involves more than science, technology, engineering and math. As computers behave more like humans, the social sciences and humanities will become even more important. Languages, art, history, economics, ethics, philosophy, psychology and human development courses can teach critical, philosophical and ethics-based skills that will be instrumental in the development and management of AI solutions. If AI is to reach its potential in serving humans, then every engineer will need to learn more about the liberal arts and every liberal arts major will need to learn more about engineering.

We're all going to need to spend more time talking with, listening to, and learning from each other. As two people from different disciplines who've benefited from doing just that, we appreciate firsthand the valuable and even enjoyable opportunities this can create.

We hope that the pages that follow can help as we all get started.

Brad Smith President and Chief Legal Officer
Harry Shum Executive Vice President, Artificial Intelligence and Research

Microsoft Corporation

1. See Brad Smith and Carol Ann Browne, "Today in Technology: The Day the Horse Lost its Job," at https://www.linkedin.com/pulse/today-technology-day-horse-lost-its-job-brad-smith/
2. Lendol Calder, Financing the American Dream: A Cultural History of Consumer Credit (Princeton: Princeton University Press, 1999), p. 184.
3. John Steele Gordon, An Empire of Wealth: The Epic History of American Economic Power (New York: HarperCollins Publishers, 2004), p. 299-300.

Brad Smith
Harry Shum

ACKNOWLEDGEMENTS

We would like to thank the following contributors for providing their insights and perspectives in the development of this book.

Benedikt Abendroth, Geff Brown, Carol Ann Browne, Dominic Carr, Pablo Chavez, Steve Clayton, Amy Colando, Jane Broom Davidson, Mariko Davidson, Paul Estes, John Galligan, Sue Glueck, Cristin Goodwin, Mary Gray, David Heiner, Merisa Heu-Weller, Eric Horvitz, Teresa Hutson, Nicole Isaac, Lucas Joppa, Aaron Kleiner, Allyson Knox, Cornelia Kutterer, Jenny Lay-Flurrie, Andrew Marshall, Anne Nergaard, Carolyn Nguyen, Barbara Olagaray, Michael Philips, Brent Sanders, Mary Snapp, Dev Stahlkopf, Steve Sweetman, Lisa Tanzi, Ana White, Joe Whittinghill, Joshua Winter, Portia Wu

Chapter 1

The Future of Artificial Intelligence

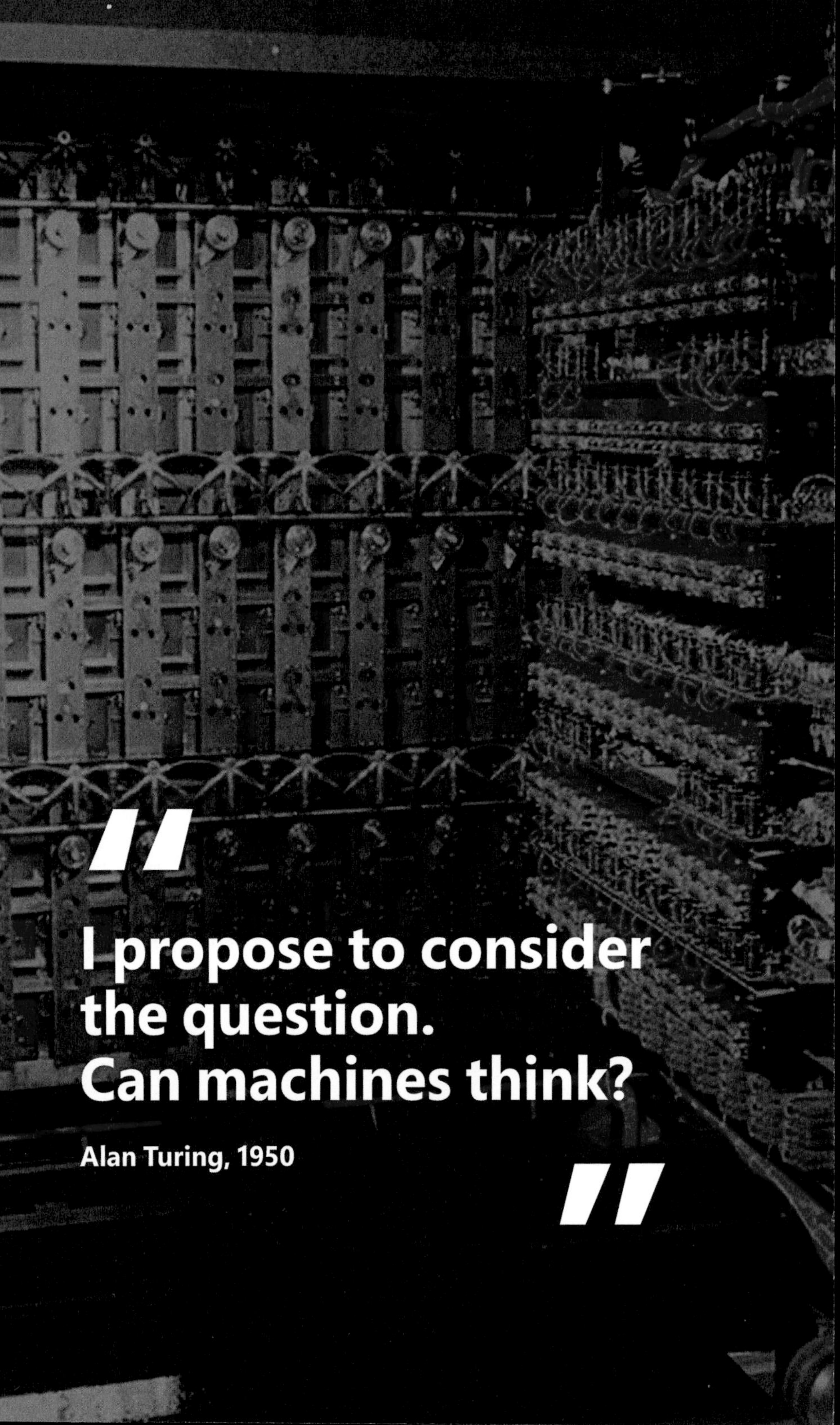
"
I propose to consider
the question.
Can machines think?
Alan Turing, 1950
"

In the summer of 1956, a team of researchers at Dartmouth College met to explore the development of computer systems capable of learning from experience, much as people do. But, even this seminal moment in the development of AI was preceded by more than a decade of exploration of the notion of machine intelligence, exemplified by Alan Turing's quintessential test: a machine could be considered "intelligent" if a person interacting with it (by text in those days) could not tell whether it was a human or a computer.

Researchers have been advancing the state of the art in AI in the decades since the Dartmouth conference. Developments in subdisciplines such as machine vision, natural language understanding, reasoning, planning and robotics have produced an ongoing stream of innovations, many of which have already become part of our daily lives. Route-planning features in navigation systems, search engines that retrieve and rank content from the vast amounts of information on the internet, and machine vision capabilities that enable postal services to automatically recognize and route handwritten addresses are all enabled by AI.

At Microsoft, we think of AI as a set of technologies that enable computers to perceive, learn, reason and assist in decision-making to solve problems in ways that are similar to what people do. With these capabilities, how computers understand and interact with the world is beginning to feel far more natural and responsive than in the past, when computers could only follow pre-programmed routines.

Not so long ago we interacted with computers via a command line interface. And while the graphical user interface was an important step forward, we will soon be routinely interacting with computers just by talking to them, just as we would to a person. To enable these new capabilities, we are, in effect, teaching computers to see, hear, understand and reason.[1] Key technologies include:

Vision: the ability of computers to "see" by recognizing what is in a picture or video.

Speech: the ability of computers to "listen" by understanding the words that people say and to transcribe them into text.

Language: the ability of computers to "comprehend" the meaning of the words, taking into account the many nuances and complexities of language (such as slang and idiomatic expressions).

Knowledge: the ability of a computer to "reason" by understanding the relationship between people, things, places, events and the like. For instance, when a search result for a movie provides information about the cast and other movies those actors were in, or at work when you participate in a meeting and the last several documents that you shared with the person you're meeting with are automatically delivered to you. These are examples of a computer reasoning by drawing conclusions about which information is related to other information.

Computers are learning the way people do; namely, through experience. For computers, experience is captured in the form of data. In predicting how bad traffic will be, for example, computers draw upon data regarding historical traffic flows based on the time of day, seasonal variations, the weather, and major events in the area such as concerts or sporting events. More broadly, rich "graphs" of information are foundational to enabling computers to develop an understanding of relevant relationships and interactions between people, entities and events. In developing AI systems, Microsoft is drawing upon graphs of information that include knowledge about the world, about work and about people.

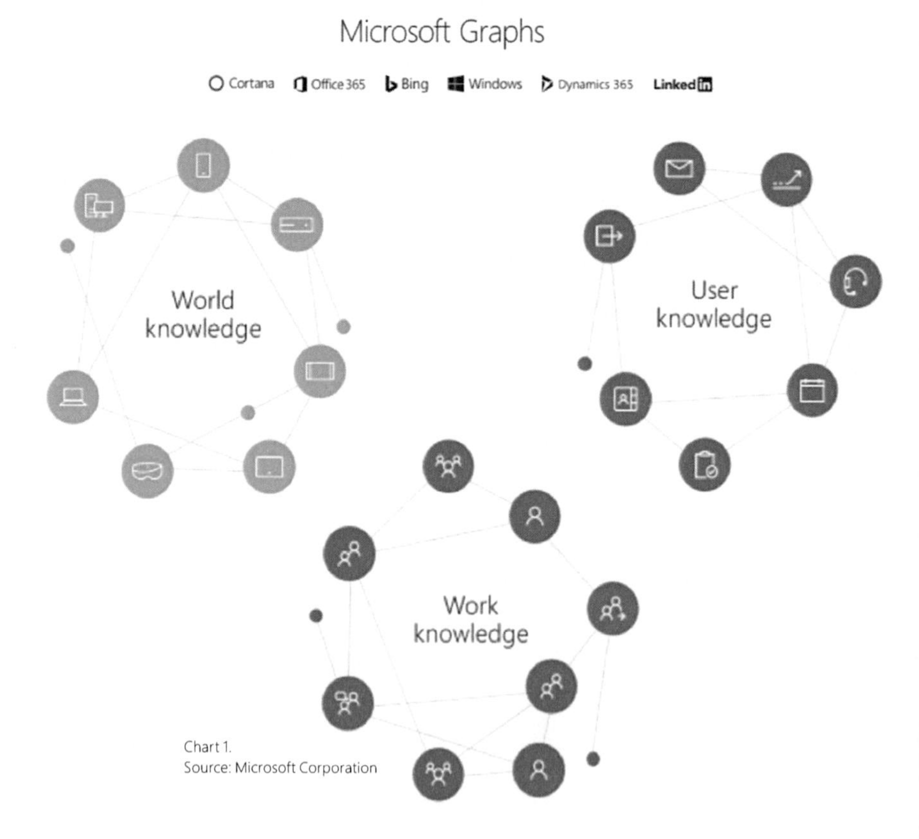

Chart 1.
Source: Microsoft Corporation

Thanks in part to the availability of much more data, researchers have made important strides in these technologies in the past few years. In 2015, researchers at Microsoft announced that they had taught computers to identify objects in a photograph or video as accurately as people do in a test using the standard ImageNet 1K database of images.[2] In 2017, Microsoft's researchers announced they had developed a speech recognition system that understood spoken words as accurately as a team of professional transcribers, with an error rate of just 5.1 percent using the standard Switchboard dataset.[3] In essence, AI-enhanced computers can, in most cases, see and hear as accurately as humans.

Much work remains to be done to make these innovations applicable to everyday use. Computers still may have a hard time understanding speech in a noisy environment where people speak over one another or when presented with unfamiliar accents or languages. It is especially challenging to teach computers to truly understand not just what words were spoken, but what the words mean and to reason by drawing conclusions and making decisions based on them. To enable computers to comprehend meaning and answer more complex questions, we need to take a big-picture view, understand and evaluate context, and bring in background knowledge.

Why Now?

Researchers have been working on AI for decades. Progress has accelerated over the past few years thanks in large part to three developments: the increased availability of data; growing cloud computing power; and more powerful algorithms developed by AI researchers.

As our lives have become increasingly digitized and sensors have become cheap and ubiquitous, more data than ever before is available for computers to learn from.

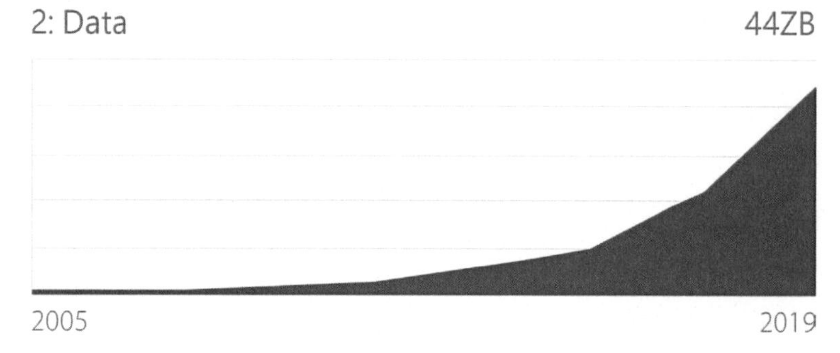

Chart 2.
Source: IDC Digital Universe Forecast, 2014

Only with data can computers discern the patterns, often subtle, that enable them to "see," "hear" and "understand."

Analyzing all this data requires massive computing power, which is available thanks to the efficiencies of cloud computing. Today, organizations of any type can tap into the power of the cloud to develop and run their AI systems.

Researchers at Microsoft, other technology firms, universities and governments have drawn upon this combination of the availability of this data, and with it ready access to powerful computing and breakthroughs in AI techniques — such as "deep learning" using so-called "deep neural nets"— to enable computers to mimic how people learn.

In many ways, AI is still maturing as a technology. Most of the progress to date has been in teaching computers to perform narrow tasks — play a game, recognize an image, predict traffic. We have a long way to go to imbue computers with "general" intelligence. Today's AI cannot yet begin to compete with a child's ability to understand and interact with the world using senses such as touch, sight and smell. And AI systems have only the most rudimentary ability to understand human expression, tone, emotion and the subtleties of human interaction. In other words, AI today is strong on "IQ" but weak on "EQ."

At Microsoft, we're working toward endowing computers with more nuanced capabilities. We believe an integrated approach that combines various AI disciplines will lead to the development of more sophisticated tools that can help people perform more complex, multifaceted tasks. Then, as we learn how to combine multiple IQ functions with abilities that come naturally to people — like applying knowledge of one task to another, having a commonsense understanding of the world, interacting naturally, or knowing when someone is trying to be funny or sarcastic, and the difference between those — AI will become even more helpful. While this is clearly a formidable challenge, when machines can integrate the smarts of IQ and the empathy of EQ in their interactions, we will have achieved what we call "conversational AI." This will be an important step forward in the evolution of computer-human interaction.

Microsoft's Approach to AI

When Bill Gates and Paul Allen founded Microsoft over 40 years ago, their aim was to bring the benefits of computing — then largely locked up in mainframes — to everyone. They set out to build a "personal" computer that would help people be more productive at home, at school and at work. Today, Microsoft is aiming to do much the same with AI. We're building AI systems that are designed to amplify natural human ingenuity. We're deploying AI systems with the goal of making them available to everyone and aspiring to build AI systems that reflect timeless societal values so that AI earns the trust of all.[4]

Amplifying Human Ingenuity

We believe that AI offers incredible opportunities to drive widespread economic and social progress. The key to attaining these benefits is to develop AI in such a way that it is human-centered. Put simply, we aim to develop AI in order to augment human abilities, especially humankind's innate ingenuity. We want to combine the capabilities of computers with human capabilities to enable people to achieve more.

Computers are very good at remembering things. Absent a system failure, computers never forget. Computers are very good at probabilistic reasoning, something many people are not so good at. Computers are very good at discerning patterns in data that are too subtle for people to notice. With these capabilities, computers can help us make better decisions. And this is a real benefit, because, as researchers in cognitive psychology have established, human decision-making is often imperfect. Broadly speaking, the kind of "computational intelligence" that computers can provide will have a significant impact in almost any field where intelligence itself has a role to play.

AI improving medical image analysis for clinicians

AI systems are already helping people tackle big problems. A good example of this is "InnerEye," a project in which U.K.-based researchers at Microsoft have teamed up with oncologists to develop an AI system to help treat cancer more effectively.[5]

InnerEye uses AI technology originally developed for video gameplay to analyze computed tomography (CT) and magnetic resonance imaging (MRI), and helps oncologists target cancer treatment more quickly. CT and MRI scans allow doctors to look inside a patient's body in three dimensions and study anomalies, such as tumors. For cancer patients who are undergoing radiation therapy, oncologists use such scans to delineate tumors from the surrounding healthy tissue, bone and organs. In turn, this helps focus the cell-damaging radiation treatment on the tumor while avoiding healthy anatomy as much as possible. Today, this 3-D delineation task is manual, slow and error-prone. It requires a radiation oncologist to draw contours on hundreds of cross-sectional images by hand, one at a time — a process that can take hours. InnerEye is being designed to accomplish the same task in a fraction of that time, while giving oncologists full control over the accuracy of the final delineation.

To create InnerEye's automatic segmentation, researchers used hundreds of raw CT and MRI scans (with all identifying patient information removed). The scans were fed into an AI system that learned to recognize tumors and healthy anatomical structures with a clinical level of accuracy. As part of the process, once the InnerEye automatic segmentation is complete, the oncologist goes in to fine-tune the contours. The doctor is in control at all times. With further advances, InnerEye may be helpful for measuring and tracking tumor changes over time, and even assessing whether a treatment is working.

AI helping researchers prevent disease outbreaks

Another interesting example is "Project Premonition." We've all seen the heartbreaking stories of lives lost in recent years to dangerous diseases like Zika, Ebola and dengue that are transmitted from animals and insects to people. Today, epidemiologists often don't learn about the emergence of these pathogens until an outbreak is underway. But this project — developed by scientists and engineers at Microsoft Research, the University of Pittsburgh, the University of California Riverside and Vanderbilt University — is exploring ways to detect pathogens in the environment so public health officials can protect people from transmission before an outbreak begins.[6]

What epidemiologists need are sensors that can detect when pathogens are present. The researchers on this project hit upon an ingenious idea: why not use mosquitoes as sensors? There are plenty of them and they feed on a wide range of animals, extracting a small amount of blood that contains genetic information about the animal bitten and pathogens circulating in the environment.

The researchers use advanced autonomous drones capable of navigating through complex environments to identify areas where mosquitoes breed. They then deploy robotic traps that can distinguish between the types of mosquitoes researchers want to collect and other insects, based on wing movement patterns. Once specimens are collected, cloud-scale genomics and advanced AI systems identify the animals that the mosquitoes have fed on and the pathogens that the animals carry. In the past, this kind of genetic analysis could take a month; now the AI capabilities of Project Premonition have shortened that to about 12 hours.

During a Zika outbreak in 2016, Project Premonition drones and traps were tested in Houston. More than 20,000 mosquitoes were collected from nine different species, including those known to carry Zika, dengue, West Nile virus and malaria. Because the traps also gather data on environmental conditions when an insect is collected, the test provided useful data not only about pathogens in the environment but also about mosquito behavior. This helped Project Premonition researchers improve their ability to target hotspots where mosquitoes breed. Researchers are also working to improve how to identify known diseases and detect the presence of previously unknown pathogens.

While the project is still in its early stages, it may well point the way toward an effective early warning system that will detect some of the world's most dangerous diseases in the environment and help prevent deadly outbreaks.

Making Human-Centered AI Available to All

We cannot deliver on the promise of AI unless we make it broadly available to all. People around the world can benefit from AI — but only if AI technologies are available for them. For Microsoft, this begins with basic R&D. Microsoft Research, with its 26-year history, has established itself as one of the premier research organizations in the world contributing both to the advancement of computer science and to Microsoft products and services. Our researchers have published more than 22,000 papers in all areas of study — from the environment to health, and from privacy to security. Recently, we announced the creation of Microsoft Artificial Intelligence and Research, a new group that brings together approximately 7,500 computer scientists, researchers and engineers. This group is chartered with pursuing a deeper understanding of the computational foundations of intelligence, and is focused on integrating research from all fields of AI research in order to solve some of AI's most difficult challenges.

We continue to encourage researchers to publish their results broadly so that AI researchers around the world — at universities, at other companies and in government settings — can build on these advances.

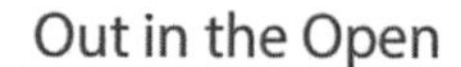

Artificial-intelligence-related research*
By company-affiliation, 2000-16

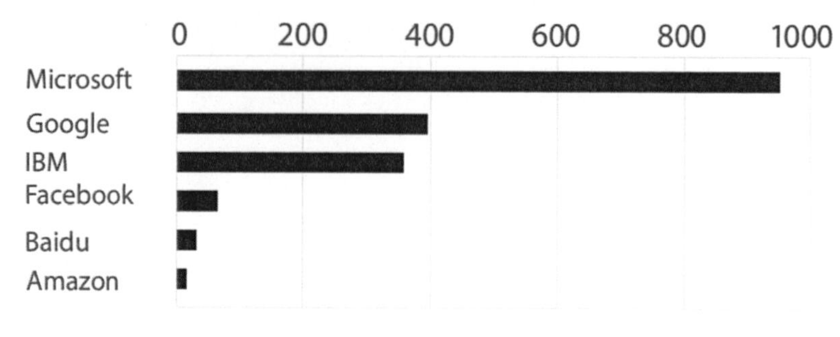

*Papers from five major AI conferences

Chart 3.
Source: The Economist

For our customers, we're building AI capabilities into our most popular products, such as Windows and Office. Windows is more secure thanks to AI systems that detect malware and automatically protect computers against it. In Office, Researcher for Word helps you write more compelling documents. Without leaving a document, you can find and incorporate relevant information from across the web using Bing "Knowledge Graph." If you are creating a PowerPoint presentation, PowerPoint Designer assesses the images and text you've used, and provides design tips to create more professional-looking slides, along with suggestions for text captions for images to improve accessibility. And PowerPoint Presentation Translator lets you engage diverse audiences more effectively by breaking down language barriers through auto-captioning in over 60 languages. This feature will also aid people with hearing loss.

AI is the enabling technology behind Cortana, Microsoft's personal digital assistant. Cortana is young, but she's learning fast. Already Cortana can help you schedule a meeting, make a restaurant reservation and find answers to questions on a broad range of topics. Over time, Cortana will be able to interact with other personal digital assistants to automatically handle tasks that take up time and follow familiar patterns. One of the key technologies that Cortana builds upon is Bing, our search service. But instead of just providing links to relevant information, Cortana uses Bing to discover answers to your questions and provide them in a variety of more context-rich ways.[7]

Microsoft is not only using AI technologies to create and enhance our own products, we are also making them available to developers so that they can build their own AI-powered products. The Microsoft AI Platform offers services, tools and infrastructure making AI development easier for developers and organizations of any size. Our service offerings include Microsoft Cognitive Services, a set of pre-built AI capabilities including vision, speech, language and search. All of these are hosted in the cloud and can be easily integrated into applications. Some of these are also customizable so that they can be better optimized to help transform and improve business processes specific to an organization's industry and business needs. You can see the breadth of these offerings below.

We also have technologies available to simplify the creation of "bots" that can engage with people more naturally and conversationally. We offer a growing collection of coding and management tools to make the AI development process

Chart 4.
Source: Microsoft Corporation

easier. And our infrastructure offerings help others develop and deploy algorithms, and store their data and derive insights from it.

Finally, with Microsoft's AI Business Solutions, we are building systems of intelligence so organizations can better understand and act on the information they collect in order to be more productive.

One example of an AI Business Solution is Customer Care Intelligence, currently being used by the Department of Human Services (DHS) in Australia to transform how it delivers services to citizens. At the heart of the program is an expert system that uses a virtual assistant named "Roxy" who helps claims processing officers answer questions and solve problems. Roxy was trained using the DHS operational blueprint that includes all of the agency's policies and procedures, and fed all of the questions that passed between claims officers and DHS managers over a three-month period. In early use, the system was able to answer nearly 80 percent of the questions it was asked. This is expected to translate to about a 20 percent reduction in workload for claims officers.

The internal project with Roxy was so successful that DHS is now developing virtual assistants that will interact directly with citizens. One of these projects will target high school seniors to help them decide whether to apply for a university or enroll in a vocational program through Australia's Technical and Further Education program by helping them navigate the qualification process.

The Potential of Modern AI – Addressing Societal Challenges

At Microsoft, we aim to develop AI systems that will enable people worldwide to more effectively address local and global challenges, and to help drive progress and economic opportunity.

Today's AI enables faster and more profound progress in nearly every field of human endeavor, and it is essential to enabling the digital transformation that is at the heart of worldwide economic development. Every aspect of a business or organization — from engaging with customers to transforming products, optimizing operations and empowering employees — can benefit from this digital transformation.

But even more importantly, AI has the potential to help society overcome some of its most daunting challenges. Think of the most complex and pressing issues that humanity faces: from reducing poverty and improving education, to delivering healthcare and eradicating diseases, addressing sustainability challenges such as growing enough food to feed our fast-growing global population through to advancing inclusion in our society. Then imagine what it would mean in lives saved, suffering alleviated and human potential unleashed if we could harness AI to help us find solutions to these challenges.

Providing effective healthcare at a reasonable cost to the approximately 7.5 billion people on the planet is one of society's most pressing challenges. Whether it's analyzing massive amounts of patient data to uncover hidden patterns that can point the way toward better treatments, identifying compounds that show promise as new drugs or vaccines, or unlocking the potential of personal medicine based on in-depth genetic analysis, AI offers vast opportunities to transform how we understand disease and improve health. Machine reading can help doctors quickly find important information amid thousands of documents that they otherwise wouldn't have time to read. By doing so, it can help medical professionals spend more of their time on higher value and potentially lifesaving work.

Providing safe and efficient transportation is another critical challenge where AI can play an important role. AI-controlled driverless vehicles could reduce traffic accidents and expand the capacity of existing road infrastructure, saving hundreds of thousands of lives every year while improving traffic flow and reducing carbon emissions. These vehicles will also facilitate greater inclusiveness in society by enhancing the independence of those who otherwise are not able to drive themselves.

In education, the ability to analyze how people acquire knowledge and then use that information to develop predictive models for engagement and comprehension points the way toward new approaches to education that combine online and teacher-led instruction and may revolutionize how people learn.

As demonstrated by Australia's Department of Human Services' use of the natural language capabilities of Customer Care Intelligence to answer questions, AI also has the potential to improve how governments interact with their citizens and deliver services.

AI enabling people with low vision to hear information about the world around them

Another area where AI has the potential to have a significant positive impact is in serving the more than 1 billion people in the world with disabilties. One example of how AI can make a difference is a recent Microsoft offering called "Seeing AI," available in the iOS app store, that can assist people with blindness and low vision as they navigate daily life.

Seeing AI was developed by a team that included a Microsoft engineer who lost his sight at 7 years of age. This powerful app, while still in its early stages, demonstrates the potential

for AI to empower people with disabilities by capturing images from the user's surroundings and instantly describing what is happening. For example, it can read signs and menus, recognize products through barcodes, interpret handwriting, count currency, describe scenes and objects in the vicinity, or, during a meeting, tell the user that there is a man and a woman sitting across the table who are smiling and paying close attention.[8]

AI empowering farmers to be more productive and increase their yield

And with the world's population expected to grow by nearly 2.5 billion people over the next quarter century, AI offers significant opportunities to increase food production by improving agricultural yield and reducing waste. For

example, our "FarmBeats" project uses advanced technology, existing connectivity infrastructure, and the power of the cloud and machine learning to enable data-driven farming at low cost. This initiative provides farmers with easily interpretable insights to help them improve agricultural yield, lower overall costs and reduce the environmental impact of farming.[9]

Given the significant benefits that stem from using AI — empowering us all to accomplish more by being more productive and efficient, driving better business outcomes, delivering more effective government services and helping to solve difficult societal issues — it's vital that everyone has the opportunity to use it. Making AI available to all people and organizations is foundational to enabling everyone to capitalize on the opportunities AI presents and share in the benefits it delivers.

The Challenges AI Presents

As with the great advances of the past on which it builds — including electricity, the telephone and transistors — AI will bring about vast changes, some of which are hard to imagine today. And, as was the case with these previous significant technological advances, we'll need to be thoughtful about how we address the societal issues that these changes bring about. Most importantly, we all need to work together to ensure that AI is developed in a responsible manner so that people will trust it and deploy it broadly, both to increase business and personal productivity and to help solve societal problems.

This will require a shared understanding of the ethical and societal implication of these new technologies. This, in turn, will help pave the way toward a common framework of principles to guide researchers and developers as they deliver a new generation of AI-enabled systems and capabilities, and governments as they consider a new generation of rules and regulations to protect the safety and privacy of citizens and ensure that the benefits of AI are broadly accessible.

In Chapter 2, we offer our initial thinking on how to move forward in a way that respects universal values and addresses the full range of societal issues that AI will raise, while ensuring that we achieve the full potential of AI to create opportunities and improve lives.

Chapter 2

Principles, Policies and Laws for the Responsible Use of AI

"

In a sense, artificial intelligence will be the ultimate tool because it will help us build all possible tools.

K. Eric Drexler

"

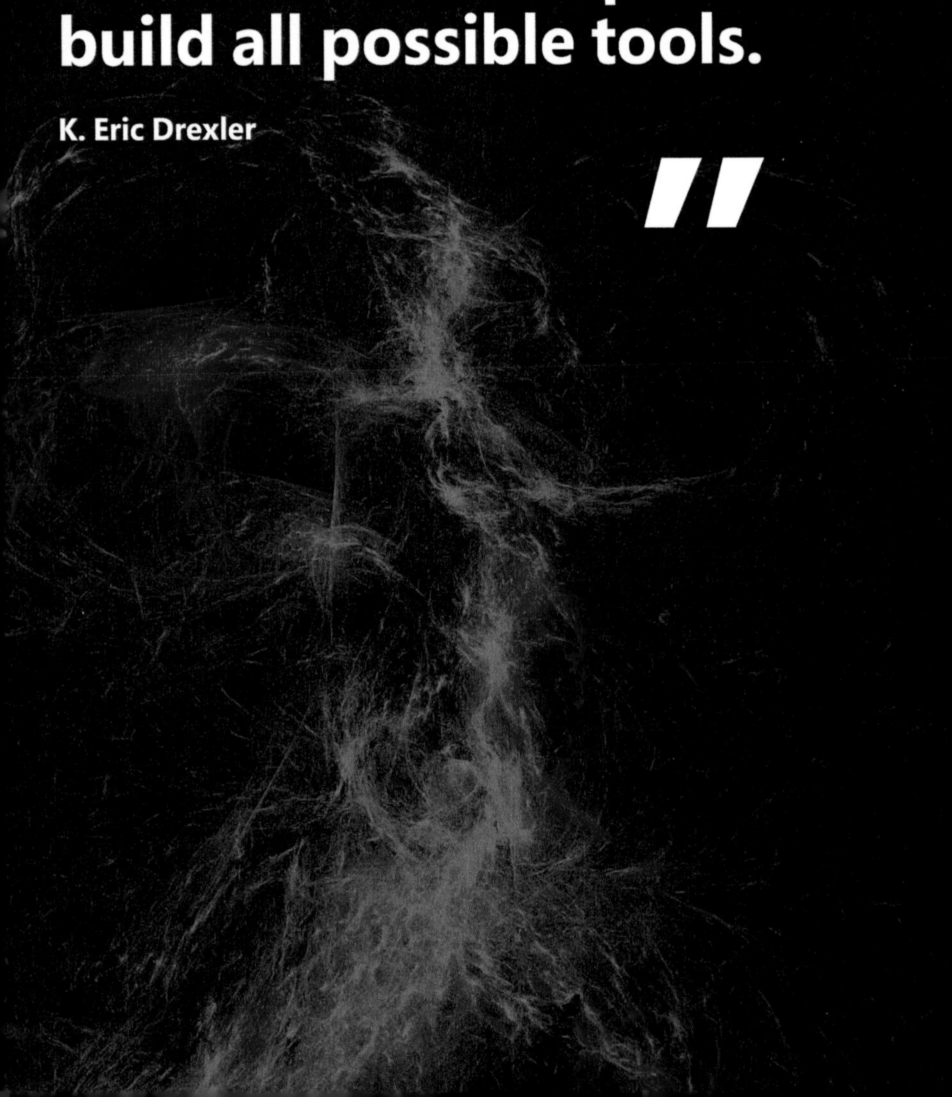

As AI begins to augment human understanding and decision-making in fields like education, healthcare, transportation, agriculture, energy and manufacturing, it will raise new societal questions. How can we ensure that AI treats everyone fairly? How can we best ensure that AI is safe and reliable? How can we attain the benefits of AI while protecting privacy? How do we not lose control of our machines as they become increasingly intelligent and powerful?

The people who are building AI systems are, of course, required to comply with the broad range of laws around the world that already govern fairness, privacy, injuries resulting from unreasonable behaviors and the like. There are no exceptions to these laws for AI systems. But we still need to develop and adopt clear principles to guide the people building, using and applying AI systems. Industry groups and others should build off these principles to create detailed best practices for key aspects of the development of AI systems, such as the nature of the data used to train AI systems, the analytical techniques deployed, and how the results of AI systems are explained to people using those systems.

It's imperative that we get this right if we're going to prevent mistakes. Otherwise people may not fully trust AI systems. And if people don't trust AI systems, they will be less likely to contribute to the development of such systems and to use them.

Ethical and Societal Implications

Business leaders, policymakers, researchers, academics and representatives of nongovernmental groups must work together to ensure that AI-based technologies are designed and deployed in a manner that will earn the trust of the people who use them and the individuals whose data is being collected. The Partnership on AI (PAI), an organization co-founded by Microsoft, is one vehicle for advancing these discussions. Important work is also underway at many universities and governmental and non-governmental organizations.[10]

Designing AI to be trustworthy requires creating solutions that reflect ethical principles that are deeply rooted in important and timeless values. As we've thought about it, we've focused on six principles that we believe should guide the development of AI. Specifically, AI systems should be fair, reliable and safe, private and secure, inclusive, transparent, and accountable. These principles are critical to addressing the societal impacts of AI and building trust as the technology becomes more and more a part of the products and services that people use at work and at home every day.

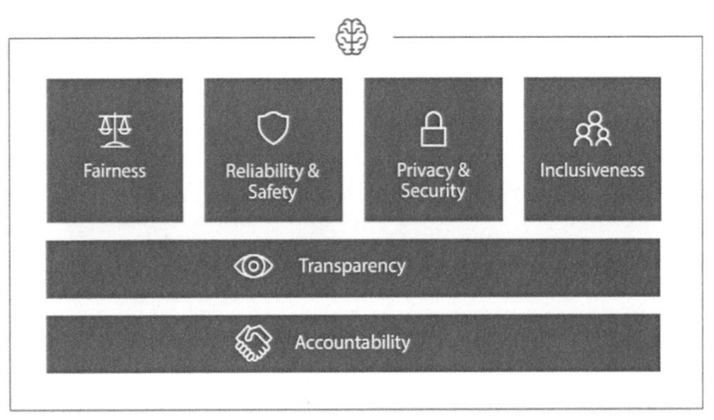

Chart 5.
Source: Microsoft Corporation

Fairness – AI systems should treat all people fairly.

AI systems should treat everyone in a fair and balanced manner and not affect similarly situated groups of people in different ways. For example, when AI systems provide guidance on medical treatment, loan applications or employment, they should make the same recommendations for everyone with similar symptoms, financial circumstances or professional qualifications. If designed properly, AI can help make decisions that are fairer because computers are purely logical and, in theory, are not subject to the conscious and unconscious biases that inevitably influence human decision-making. Yet, because AI systems are designed by human beings and the systems are trained using data that reflects the imperfect world in which we live, AI can operate unfairly without careful planning. To ensure that fairness is the foundation for solutions using this new technology, it's imperative that developers understand how bias can be introduced into AI systems and how it can affect AI-based recommendations.

The design of any AI systems starts with the choice of training data, which is the first place where unfairness can arise. Training data should sufficiently represent the world in which we live, or at least the part of the world where the AI system will operate. Consider an AI system that enables facial recognition or emotion detection. If it is trained solely on images of adult faces, it may not accurately identify the features or expressions of children due to differences in facial structure.

But ensuring the "representativeness" of data is not enough. Racism and sexism can also creep into societal data. Training an AI system on such data may inadvertently lead to results that perpetuate these harmful biases. One example might be an AI system designed to help employers screen job applicants. When trained on data from public employment records, this system might "learn" that most software developers are male. As a result, it may favor men over women when selecting candidates for software developer positions, even though the company deploying the system is seeking to promote diversity through its hiring practices.[11]

An AI system could also be unfair if people do not understand the limitations of the system, especially if they assume technical systems are more accurate and precise than people, and therefore more authoritative. In many cases, the output of an AI system is actually a prediction. One example might be "there is a 70 percent likelihood that the applicant will default on the loan." The AI system may be highly accurate, meaning that if the bank extends credit every time to people with the 70 percent "risk of default," 70 percent of those people will, in fact, default. Such a system may be unfair in application, however, if loan officers incorrectly interpret "70 percent risk of default" to simply mean "bad credit risk" and decline to extend credit to everyone with that score — even though nearly a third of those applicants are predicted to be a good credit risk. It will be essential to train people to understand the meaning and implications of AI results to supplement their decision-making with sound human judgment.

How can we ensure that AI systems treat everyone fairly? There's almost certainly a lot of learning ahead for all of us in this area, and it will be vital to sustain research and foster robust discussions to share new best practices that emerge. But already some important themes are emerging.

First, we believe that the people designing AI systems should reflect the diversity of the world in which we live. We also believe that people with relevant subject matter expertise (such as those with consumer credit expertise for a credit scoring AI system) should be included in the design process and in deployment decisions.

Second, if the recommendations or predictions of AI systems are used to help inform consequential decisions about people, we believe it will be critical that people are primarily accountable for these decisions. It will also be important to invest in research to better understand the impact of AI systems on human decision-making generally.

Finally — and this is vital — industry and academia should continue the promising work underway to develop analytical techniques to detect and address potential unfairness, like methods that systematically assess the data used to train AI systems for appropriate representativeness and document information about its origins and characteristics.

Ultimately, determining the full range of work needed to address possible bias in AI systems will require ongoing discussions that include a wide range of interested stakeholders. Academic research efforts such as those highlighted at the annual conference for researchers on Fairness, Accountability, and Transparency in Machine Learning have raised awareness of the issue. We encourage increased efforts across the public, private and civil sectors to expand these discussions to help find solutions.

Reliability – AI systems should perform reliably and safely.

The complexity of AI technologies has fueled fears that AI systems may cause harm in the face of unforeseen circumstances, or that they can be manipulated to act in harmful ways. As is true for any technology, trust will ultimately depend on whether AI-based systems can be operated reliably, safely and consistently — not only under normal circumstances but also in unexpected conditions or when they are under attack.

This begins by demonstrating that systems are designed to operate within a clear set of parameters under expected performance conditions, and that there is a way to verify that they are behaving as intended under actual operating conditions. Because AI systems are data-driven, how they behave and the variety of conditions they can handle reliably and safely largely reflects the range of situations and circumstance that developers anticipate during design and testing. For example, an AI system designed to detect misplaced objects may have difficulty recognizing items in low lighting conditions, meaning designers should conduct tests in typical and poorly lit environments. Rigorous testing is essential during system development and deployment to ensure that systems can respond safely to unanticipated situations; do not have unexpected performance failures; and do not evolve in ways that are inconsistent with original expectations.

Design and testing should also anticipate and protect against the potential for unintended system interactions or bad actors to influence operations, such as through cyberattacks. Securing AI systems will require developers to identify abnormal behaviors and prevent manipulation, such as the introduction of malicious data that may be intended to negatively impact AI behavior.

In addition, because AI should augment and amplify human capabilities, people should play a critical role in making decisions about how and when an AI system is deployed, and whether it's appropriate to continue to use it over time. Since AI systems often do not see or understand the bigger societal picture, human judgment will be key to identifying potential blind spots and biases in AI systems. Developers should be cognizant of these challenges as they build and deploy systems, and share information with their customers to help them monitor and understand system behavior so that they can quickly identify and correct any unintended behaviors that may surface.

In one example in the field of AI research, a system designed to help make decisions about whether to hospitalize patients with pneumonia "learned" that people with asthma have a lower rate of mortality from pneumonia than the general population. This was a surprising result because people with asthma are generally considered to be at greater risk of dying from pneumonia than others. While the correlation was accurate, the system failed to detect that the primary reason for this lower mortality rate was that asthma patients receive faster and more comprehensive care than other patients because they are at greater risk. If researchers hadn't noticed that the AI system had drawn a misleading inference, the system might have recommended against hospitalizing people with asthma, an outcome that would have run counter to what the data revealed.[12] This highlights the critical role that people, particularly those with subject matter expertise, must play in observing and evaluating AI systems as they are developed and deployed.

Principles of robust and fail-safe design that were pioneered in other engineering disciplines can be valuable in designing and developing reliable and safe AI systems. Research and collaboration involving industry participants, governments, academics and other experts to further improve the safety and reliability of AI systems will be increasingly important as AI systems become more widely used in fields such as transportation, healthcare and financial services.

We believe the following steps will promote the safety and reliability of AI systems:

- Systematic evaluation of the quality and suitability of the data and models used to train and operate AI-based products and services, and systematic sharing of information about potential inadequacies in training data.
- Processes for documenting and auditing operations of AI systems to aid in understanding ongoing performance monitoring.
- When AI systems are used to make consequential decisions about people, a requirement to provide adequate explanations of overall system operation, including information about the training data and

algorithms, training failures that have occurred, and the inferences and significant predictions generated, especially.

- Involvement of domain experts in the design process and operation of AI systems used to make consequential decisions about people.
- Evaluation of when and how an AI system should seek human input during critical situations, and how a system controlled by AI should transfer control to a human in a manner that is meaningful and intelligible.
- A robust feedback mechanism so that users can easily report performance issues they encounter.

Creating AI systems that are safe and reliable is a shared responsibility. It is, therefore, critically important for industry participants to share best practices for design and development, such as effective testing, the structure of trials and reporting. Topics such as human-robot interaction and how AI-driven systems that fail should hand control over to people are important areas not only for ongoing research, but also for enhanced collaboration and communication within the industry.

Privacy & Security – AI systems should be secure and respect privacy.

As more and more of our lives are captured in digital form, the question of how to preserve our privacy and secure our personal data is becoming more important and more complicated. While protecting privacy and security is important to all technology development, recent advances require that we pay even closer attention to these issues to create the levels of trust needed to realize the full benefits of AI. Simply put, people will not share data about themselves — data that is essential for AI to help inform decisions about people — unless they are confident that their privacy is protected and their data secured.

Privacy needs to be both a business imperative and a key pillar of trust in all cloud computing initiatives. This is why Microsoft made firm commitments to protect the security and privacy of our customers' data, and why we are upgrading our engineering systems to ensure that we satisfy data protection laws around the world, including the European Union's General Data Protection Regulation (GDPR). Microsoft is investing in the infrastructure and systems to enable GDPR compliance in our largest-ever engineering effort devoted to complying with a regulatory environment.

Like other cloud technologies, AI systems must comply with privacy laws that require transparency about the collection, use and storage of data, and mandate that consumers have appropriate controls so that they can choose how their data is used. AI systems should also be designed so that private information is used in accordance with privacy standards and protected from bad actors who might seek to steal private information or inflict harm. Industry processes should be developed and implemented for the following: tracking relevant information about customer data (such as when it was collected and the terms governing its collection); accessing and using that data; and auditing access and use. Microsoft is continuing to invest in robust compliance technologies and processes to ensure that data collected and used by our AI systems is handled responsibly.

What is needed is an approach that promotes the development of technologies and policies that protect privacy while facilitating access to the data that AI systems require to operate effectively. Microsoft has been a leader in creating and advancing innovative state-of-the-art techniques for protecting privacy, such as differential privacy, homomorphic encryption, and techniques to separate data from identifying information about individuals and for protecting against misuse, hacking or tampering. We believe these techniques will reduce the risk of privacy intrusions by AI systems so they can use personal data without accessing or knowing the identities of individuals. Microsoft will continue to invest in research and work with governments and others in industry to develop effective and efficient privacy protection technologies that can be deployed based on the sensitivity and proposed uses of the data.

Inclusiveness – AI systems should empower everyone and engage people.

If we are to ensure that AI technologies benefit and empower everyone, they must incorporate and address a broad range of human needs and experiences. Inclusive design practices will help system developers understand and address potential barriers in a product or environment that could unintentionally exclude people. This means that AI systems should be designed to understand the context, needs and expectations of the people who use them.

The importance that information and communications technology plays in the lives of the 1 billion people around the world with disabilities is broadly recognized. More than 160 countries have ratified the United Nations Convention on the Rights of Persons with Disabilities, which covers access to digital technology in education and employment. In the United States, the Americans with Disabilities Act and the Communications and Video Accessibility Act require technology solutions to be accessible, and federal and state regulations mandate the procurement of accessible technology, as does European Union law. AI can be a powerful tool for increasing access to information, education, employment, government services, and social and economic opportunities. Real-time speech-to-text transcription, visual recognition services, and predictive text functionality that suggests words as people type are just a few examples of AI-enabled services that are already empowering those with hearing, visual and other impairments.

We also believe that AI experiences can have the greatest positive impact when they offer both emotional intelligence and cognitive intelligence, a balance that can improve predictability and comprehension. AI-based personal agents, for example, can exhibit user awareness by confirming and, as necessary, correcting understanding of the user's intent, and by recognizing and adjusting to the people, places and events that are most important to users. Personal agents should provide information and make recommendations in ways that are contextual and expected. They should provide information that helps people understand what inferences the system is making about them. Over time, such successful interactions will increase usage of AI system and trust in their performance.

Transparency – AI systems should be understandable.

Underlying these four preceding values are two foundational principles that are essential for ensuring the effectiveness of the rest: transparency and accountability.

When AI systems are used to help make decisions that impact people's lives, it is particularly important that people understand how those decisions were made. An approach that is most likely to engender trust with users and those affected by these systems is to provide explanations that include contextual information about how an AI system works and interacts with data. Such information will make it easier to identify and raise awareness of potential bias, errors and unintended outcomes.

Simply publishing the algorithms underlying AI systems will rarely provide meaningful transparency. With the latest (and often most promising) AI techniques, such as deep neural networks, there typically isn't any algorithmic output that would help people understand the subtle patterns that systems find. This is why we need a more holistic approach in which AI system designers describe the key elements of the system as completely and clearly as possible.

Microsoft is working with the Partnership on AI and other organizations to develop best practices for enabling meaningful transparency of AI systems. This includes the practices described above and a variety of other methods, such as an approach to determine if it's possible to use an algorithm or model that is easier to understand in place of one that is more complex and difficult to explain. This is an area that will require further research to understand how machine learning models work and to develop new techniques that provide more meaningful transparency.

Accountability

Finally, as with other technologies and products, the people who design and deploy AI systems must be accountable for how their systems operate. To establish accountability norms for AI, we should draw upon experience and practices in other areas, including healthcare and privacy. Those who develop and use AI systems should consider such practices and periodically check whether they are being adhered to and if they are working effectively. Internal review boards can provide oversight and guidance on which practices should be adopted to help address the concerns discussed above, and on particularly important questions regarding development and deployment of AI systems.

Internal Oversight and Guidance – Microsoft's AI and Ethics in Engineering and Research (AETHER) Committee

Ultimately these six principles need to be integrated into ongoing operations if they're to be effective. At Microsoft, we're addressing this in part through the AI and Ethics in Engineering and Research (AETHER) Committee. This committee is a new internal organization that includes senior leaders from across Microsoft's engineering, research, consulting and legal organizations who focus on proactive formulation of internal policies and on how to respond to specific issues as they arise. The AETHER Committee considers and defines best practices, provides guiding principles to be used in the development and deployment of Microsoft's AI products and solutions, and helps resolve questions related to ethical and societal implications stemming from Microsoft's AI research, product and customer engagement efforts.

Developing Policy and Law for Artificial Intelligence

AI can serve as a catalyst for progress in almost every area of human endeavor. But, as with any innovation that pushes us beyond current knowledge and experience, the advent of AI raises important questions about the relationship between people and technology, and the impact of new technology-driven capabilities on individuals and communities.

We are the first generation to live in a world where AI will play an expansive role in our daily lives. It's safe to say that most current standards, laws and regulations were not written specifically to account for AI. But, while existing rules may not have been crafted with AI in mind, this doesn't mean that AI-based products and services are unregulated. Current laws that, for example, protect the privacy and security of personal information, that govern the flow of data and how it is used, that promote fairness in the use of consumer information, or that govern decisions on credit or employment apply broadly to digital products and services or their use in decision-making, whether they explicitly mention AI capabilities or not. AI-based services are not exempt from the requirements that will take effect with GDPR, for example, or from HIPAA regulations that protect the privacy of healthcare data in the United States, or existing regulations on automobile safety.

As the role of AI continues to grow, it will be natural for policymakers not only to monitor its impact, but to address new questions and update laws. One goal should be to ensure that governments work with businesses and other stakeholders to strike the balance that is needed to maximize the potential of AI to improve people's lives and address new challenges as they arise.

As this happens, it seems inevitable that "AI law" will emerge as an important new legal topic. But, over what period of time? And in what ways should such a field develop and evolve?

We believe the most effective regulation can be achieved by providing all stakeholders with sufficient time to identify and articulate key principles guiding the development of responsible and trustworthy AI, and to implement these principles by adopting and refining best practices. Before devising new regulations or laws, there needs to be some clarity about the fundamental issues and principles that must be addressed.

The evolution of information privacy laws in the United States and Europe offers a useful model. In 1973, the United States Department of Health, Education and Welfare (HEW) issued a comprehensive report analyzing a host of societal concerns arising from the increasing computerization of information and the growing repositories of personal data held by federal agencies.[13] The report espoused a series of important principles — the Fair Information Practices — that sought to delineate fundamental privacy ideals regardless of the specific context or technology involved. Over the ensuing decades, these principles — thanks in large part to their fundamental and universal nature — helped frame a series of federal and state laws governing the collection and use of personal information within education, healthcare, financial services and other areas. Guided by these principles, the United States Federal Trade Commission (FTC) began fashioning a body of privacy case law to prevent unfair or deceptive practices affecting commerce.

Internationally, the Fair Information Practices influenced the development of local and national laws in European jurisdictions, including Germany and France, which in many respects emerged as the leaders in the development of privacy law. Beginning in the late 1970s, the Organization for Economic Coordination and Development (OECD) built upon the Fair Information Practices to promulgate its seminal Privacy Guidelines. As with the HEW's Fair Information Practices, the universal and extensible nature of the OECD's Privacy Guidelines ultimately allowed them to serve as the building blocks for the European Union's comprehensive Data Protection Directive in 1995 and its successor, the General Data Protection Regulation.

Laws in the United States and Europe ultimately diverged, with the United States pursuing a more sectoral approach and the European Union adopting more comprehensive regulation. But, in both cases, they built on universal, foundational concepts and in some cases existing laws and legal tenets. These rules addressed a very broad range of new technologies, uses and business models, as well as an increasingly diverse set of societal needs and expectations.

Today, we believe policy discussions should focus on continued innovation and advancement of fundamental AI technologies, support the development and deployment of AI capabilities across different sectors, encourage outcomes that are aligned with a shared vision of human-centered AI, and foster the development and sharing of best practices to promote trustworthy and responsible AI. The following considerations will help policymakers craft a framework to realize these objectives.

The Importance of Data

It seems likely that many near-term AI policy and regulatory issues will focus on the collection and use of data. The development of more effective AI services requires the use of data — often as much relevant data as possible.

And yet access to and use of data also involves policy issues that range from ensuring the protection of individual privacy and the safeguarding of sensitive and proprietary information to answering a range of new competition law questions. A careful and productive balancing of these objectives will require discussion and cooperation between governments, industry participants, academic researchers and civil society.

On the one hand, we believe governments should help accelerate AI advances by promoting common approaches to making data broadly available for machine learning. A large amount of useful data resides in public datasets – data that belongs to the public itself. Governments can also invest in and promote methods and processes for linking and combining related datasets from public and private organizations while preserving confidentiality, privacy and security as circumstances require.

At the same time, it will be important for governments to develop and promote effective approaches to privacy protection that take into account the type of data and the context in which it is used. To help reduce the risk of privacy intrusions, governments should support and promote the development of techniques that enable systems to use personal data without accessing or knowing the identities of individuals. Additional research to enhance "de-identification" techniques and ongoing discussions about how to balance the risks of re-identification against the social benefits will be important.

As policymakers look to update data protection laws, they should carefully weigh the benefits that can be derived from data against important privacy interests. While some sensitive personal information, such as Social Security numbers, should typically be subject to high levels of protection, rigid approaches should be avoided because the sensitivity of personal information often depends on the context in which it is provided and used. For example, an individual's name in a company directory is not typically considered sensitive and should probably require less privacy protection than if it appeared in an adoption record. In general, updated laws should recognize that processing sensitive information may be increasingly critical to serving clear public interests such as preventing the spread of communicable diseases and other serious threats to health.

Another important policy area involves competition law. As vast amounts of data are generated through the use of smart devices, applications and cloud-based services, there are growing concerns about the concentration of information by a relatively small number of companies. But, in addition to the data that companies generate from their customers, there is publicly available data. Governments can help add to the supply of available data by ensuring that public data is usable by AI developers on a non-exclusive basis. These steps will help enable developers of all types to take greater advantage of AI technologies.

At the same time, governments should monitor whether access to unique datasets (in other words, data for which there is no substitute) is becoming a barrier to competition and needs to be addressed. Other concerns relate to whether too much data is available to too few firms and whether sophisticated algorithms will enable rivals to effectively "fix" prices. All these questions warrant attention; but, they probably can be addressed within the framework of existing competition law. The question of the availability of data will arise most directly when one firm seeks to buy another and competition authorities need to consider whether the combined firms would possess datasets that are so valuable and unique that no other firms can compete effectively. Such situations are unlikely to arise very often given the vast amount of data being generated by digital technologies, the fact that multiple firms often have the same data, and the reality that people often use multiple services that generate data for a variety of firms.

Algorithms can help increase price transparency, which will help businesses and consumers buy products at the lowest cost. But, algorithms could one day become so sophisticated that firms employing them to set prices might establish the same prices, even if the firms did not agree among themselves to do so. Competition authorities will need to carefully study the benefits of price transparency as well as the risk that transparency could over time reduce price competition.

Promoting Responsible and Effective Uses of AI

In addition to addressing issues relating to data, governments have an important role to play in promoting responsible and effective uses of AI itself. This should start with the adoption of responsible AI technologies in the public sector. While enabling more effective delivery of services for citizens, this will also provide governments with firsthand experience in developing best practices to address the ethical principles identified above.

Governments also have an important role to play in funding core research to further advance AI development and support multidisciplinary research that focuses on studying and fostering solutions to the socioeconomic issues that may arise as AI technologies are deployed. This multidisciplinary research will also be valuable for the design of future AI laws and regulations.

Governments should also stimulate adoption of AI technologies across a wide range of industries and for businesses of all sizes, with an emphasis on providing incentives for small and medium-sized organizations. Promoting economic growth and opportunity by giving smaller businesses access to the capabilities that AI methods offer can play an important role in addressing income stagnation and mitigating political and social tensions that can arise as income inequality increases. As governments take these steps, they can adopt safeguards to ensure that AI is not used to discriminate either intentionally or unintentionally in a manner prohibited under applicable laws.

Liability

Governments must also balance support for innovation with the need to ensure consumer safety by holding the makers of AI systems responsible for harm caused by unreasonable practices. Well-tested principles of negligence law are most appropriate for addressing injuries arising from the deployment and use of AI systems. This is because they encourage reasonable conduct and hold parties accountable if they fall short of that standard. This works particularly well in the context of AI for a number of reasons. First, the potential roles AI systems can play and the benefit they can bring are substantial. Second, society is already familiar with a broad range of automated systems and many other existing and prospective AI technologies and services. And third, considerable work is ongoing to help mitigate the risk of harm from these systems.

Relying on a negligence standard that is already applicable to software generally to assign responsibility for harm caused by

AI is the best way for policymakers and regulators to balance innovation and consumer safety, and promote certainty for developers and users of the technology. This will help keep firms accountable for their actions, align incentives and compensate people for harm.

Fostering Dialogue and the Sharing of Best Practices

To maximize AI's potential to deliver broad-based benefits, while mitigating risks and minimizing unintended consequences, it will be essential that we continue to convene open discussions among governments, businesses, representatives from non-governmental organizations and civil society, academic researchers, and all other interested individuals and organizations. Working together, we can identify issues that have clear societal or economic consequences and prioritize the development of solutions that protect people without unnecessarily restricting future innovation.

One helpful step we can take to address current and future issues is to develop and share innovative best practices to guide the creation and deployment of people-centered AI. Industry-led organizations such as Partnership on AI that bring together industry, nonprofit organizations and NGOs can serve as forums for the process of devising and promulgating best practices. By encouraging open and honest discussion and assisting in the sharing of best practices, governments can also help create a culture of cooperation, trust and openness among AI developers, users and the public at large. This work can serve as the foundation for future laws and regulations.

In addition it will be critical that we acknowledge the broad concerns that have been raised about the impact of these technologies on jobs and the nature of work, and take steps to ensure that people are prepared for the impact that AI will have on the workplace and the workforce. Already, AI is transforming the relationship between businesses and employees, and changing how, when and where people work. As the pace of change accelerates, new skills will be essential and new ways of connecting people to training and to jobs will be required.

In Chapter 3, we look at the impact of AI on jobs and work, and offer some suggestions for steps we can take together to provide education and training for people of every age and at every stage of school and their working lives to help them take advantage of the opportunities of the AI era. We also explore the need to rethink protections for workers and social safety net programs in a time when the relationship between workers and employers is undergoing rapid change.

Chapter 3

AI and the Future of Jobs and Work

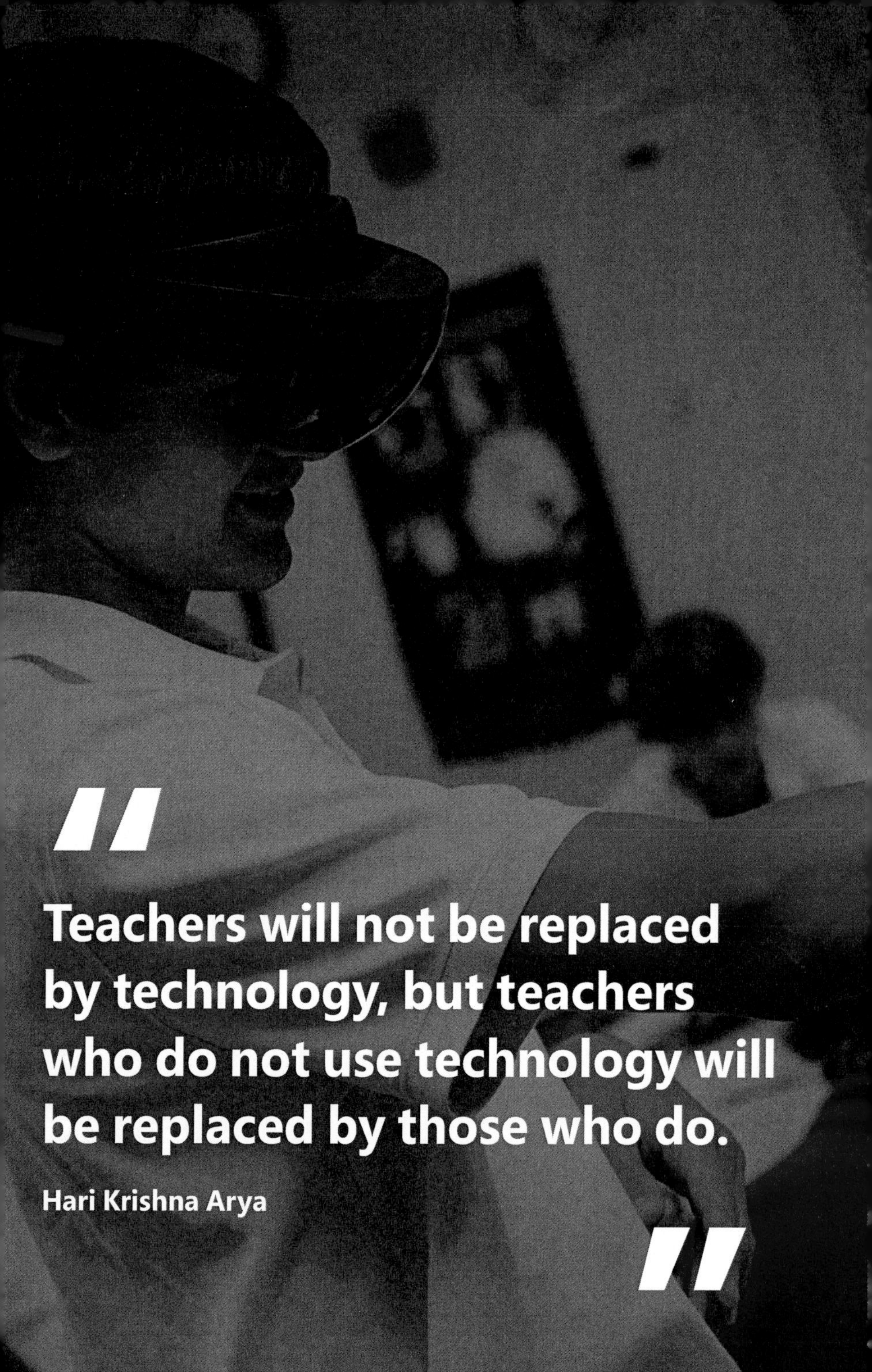
"
Teachers will not be replaced by technology, but teachers who do not use technology will be replaced by those who do.
Hari Krishna Arya
"

For more than 250 years, technology innovation has been changing the nature of jobs and work. In the 1740s, the First Industrial Revolution began moving jobs away from homes and farms to rapidly growing cities. The Second Industrial Revolution, which began in the 1870s, continued this trend, and led to the assembly line, the modern corporation, and workplaces that started to resemble offices that we would recognize today. The shift from reliance on horses to automobiles eliminated numerous occupations while creating new categories of jobs that no one initially imagined.[14] Sweeping economic changes also created difficult and sometimes dangerous working conditions that led governments to adopt labor protections and practices that are still in place today.

The Third Industrial Revolution of the past few decades created changes that many of us have experienced. For Microsoft, this was evident in how the original vision of our company — to put a computer on every desk and in every home — became reality. That transformation brought information technology into the workplace, changing how people communicate and collaborate at work, while adding new IT positions and largely eliminating jobs for secretaries who turned handwritten prose into typed copy.

Now technology is changing again the nature of jobs and work is changing with it. While available economic data is far from perfect, there are clear indications that how enterprises organize work, how people find work, and the skills that people need to prepare for work are shifting significantly. These changes are likely to accelerate in the decade ahead.

AI and cloud computing are the driving force behind much of this change. This is evident in the burgeoning on-demand — or "gig" — economy where digital platforms not only match the skills of workers with consumer or enterprise needs, they provide for people to work increasingly from anywhere in the world. AI and automation are already influencing which jobs, or aspects of jobs, will continue to exist. Some estimate that as many as 5.1 million jobs will be lost within the next decade; but, new areas of economic opportunity will also be created, as well as entirely new occupations and categories of work.[15]

These fundamental changes in the nature of work will require new ways of thinking about skills and training to ensure that workers are prepared for the future and that there is sufficient talent available for critical jobs. The education ecosystem will need to evolve as well; to help workers become lifelong learners, to enable individuals to cultivate skills that are uniquely human, and to weave ongoing education into full-time and on-demand work. For businesses, they will need to rethink how they find and evaluate talent, broaden the pool of candidates they draw from and use work portfolios to assess competence and skill. Employers will also need to focus more on offering on-the-job training, opportunities to acquire new skills, and access to outside education for their existing workforces.

In addition to rethinking how workers are trained and remain prepared for work, it is important to consider what happens to workers as traditional models of employment that typically include benefits and protections change significantly. The rapid evolution of work could undermine worker protections and benefits including unemployment insurance, workers' compensation and, in the United States, the Social Security system. To prevent this, the legal frameworks governing employment will need to be modernized to recognize new ways of working, provide adequate worker protections, and maintain the social safety net.

The Impact of Technology on Jobs and Work

Throughout history, the emergence of new technologies has been accompanied by dire warnings about human redundancy. For example, a 1928 headline in the New York Times warned that "The March of the Machine Makes Idle Hands."[16] More often, however, the reality is that new technologies have created more jobs than they destroyed. The invention of the steam engine, for example, led to the development of the steam locomotive, which was an important catalyst in the shift from a largely rural and agricultural society to one where more and more people lived in urban centers and worked in manufacturing and transportation — a transformation that changed how, when and where people worked. More recently, automated teller machines (ATMs) took over many traditional tasks for bank tellers. As a result, the average number of bank tellers per branch in the United States fell from 20 in 1988 to 13 in 2004.[17] Despite this reduction, the need for fewer tellers made it cheaper to run each branch and allowed banks to open more branches, thereby increasing the total number of employees. Instead of destroying jobs, ATMs eliminated routine tasks, which allowed bank tellers to focus on sales and customer service.[18]

This pattern is common across almost every industry. As one economist found in a recent analysis of the workforce, between 1982 and 2002, employment grew significantly faster in occupations that used computers because automation enabled workers to focus on other parts of their jobs; this increased demand for human workers to handle higher-value tasks that had not been automated.[19]

More recently, public debate has centered on the impact of automation and AI on employment. Although the terms "automation" and "AI" are often used interchangeably, the technologies are different. With automation, systems are programmed to perform specific repetitive tasks. For example, word processing automates tasks previously done by people on typewriters. Barcode scanners and point-of-sale systems automate tasks that had been done by retail employees. AI, on the other hand, is designed to seek patterns, learn from experiences, and make appropriate decisions — it does not require an explicit programmed path to determine how it will respond to the situations it encounters. Together, automation and AI are accelerating changes to the nature of jobs. As one commentator put it, "automated machines collate data — AI systems 'understand' it. We're looking at two very different systems that perfectly complement each other."[20]

700

As AI complements and accelerates automation, policymakers in countries around the world recognize that it will be an important driver of economic growth in the decades ahead. For example, China recently announced its intention to become the global leader in AI to strengthen its economy and create competitive advantages.[21]

Any business or organization that depends upon data and information — which today is almost every business and organization — can benefit from AI. These systems will improve efficiency and productivity while enabling the creation of higher-value services that can drive economic growth. But as far back as the First Industrial Revolution, the introduction of any new technology has caused concern about the impact on jobs and employment — AI and automation are no different. Indeed, it would appear that AI and automation are raising serious questions about the potential loss of jobs in developed countries. A recent survey commissioned by Microsoft found that in all 16 countries surveyed, the impact of AI on employment was identified as a key risk.[22] As machines become capable of performing tasks that require complex analysis and discretionary judgment, the concern is it will accelerate the rate of job loss beyond what already occurs due to automation.

While it's not yet clear whether AI will be more disruptive than earlier technological advances, there's no question that it is having an impact on jobs and employment. As was the case in earlier periods of significant technology transformation, it is difficult to predict how many jobs will be affected. A widely quoted University of Oxford study estimated that 47 percent of total employment in the United States is at risk due to computerization.[23] A World Bank study predicted that 57 percent of jobs in OECD countries could be automated.[24] And according to a recent paper on robots and jobs, researchers found that each robot deployed per thousand workers decreased employment by 6.2 workers and caused a decline in wages of 0.7 percent.[25]

Jobs across many industries are susceptible to the dual impact of AI and automation. Here are a few examples: a company based in San Francisco has developed "Tally" which automates the auditing of grocery store shelves to ensure goods are properly stocked and priced;[26] at Amazon, they currently use more than 100,000 robots in its fulfillment centers and is creating convenience stores with no cashiers; in Australia a company has developed a robot that can lay 1,000 bricks per hour (a task that would take human laborers a day or longer to complete); in call centers, they are using chatbots to answer customer support questions; and even in journalism, tasks such as writing summaries of sporting events are being automated.[27]

Even where jobs are not entirely replaced, AI will have an impact. In warehouses, employees have shifted from stacking bins to monitoring robots. In legal environments, paralegals and law clerks now use "e-discovery" software to find documents. In hospitals, machine learning can help doctors diagnose illnesses more quickly and enable teachers to assess student learning more effectively. But, while AI is changing these jobs, they have not disappeared; there are aspects of the work that simply cannot be automated. Many jobs will continue to require uniquely human skills that AI and

machines cannot replicate, such as creativity, collaboration, abstract and systems thinking, complex communication, and the ability to work in diverse environments.

And while it is true that AI will eliminate and change some jobs, it will also create new ones. A recent report from the research firm Forrester projects that by 2027, AI will displace 24.7 million jobs and create 14.9 million new jobs.[28] New jobs will emerge as AI changes how work is done and what people need from the world around them. Many of these jobs will be in technology. For example, banks will need network engineers instead of tellers. Retailers will need people with web programming skills to create online shopping experiences instead of greeters or salespeople on the floor. Farms will need agricultural data analysts instead of fruit pickers. Demand for data scientists, robotics experts and AI engineers will increase significantly.

What's more, AI will create jobs we cannot yet even imagine. While it is relatively easy to see where automation may reduce the need for workers, it is impossible to foresee all of the changes that will come. As one economic historian put it, "we can't predict what jobs will be created in the future, but it's always been like that."[29]

One result of the rapid transformation of work caused by AI and automation is a shortage of critical talent across many industries. As jobs increasingly require technology skills, companies compete for the employees who have specialized skills supporting digital capabilities such as robotics, augmented reality computations, cybersecurity and data science. It is estimated that by 2020, 30 percent of

DANGER

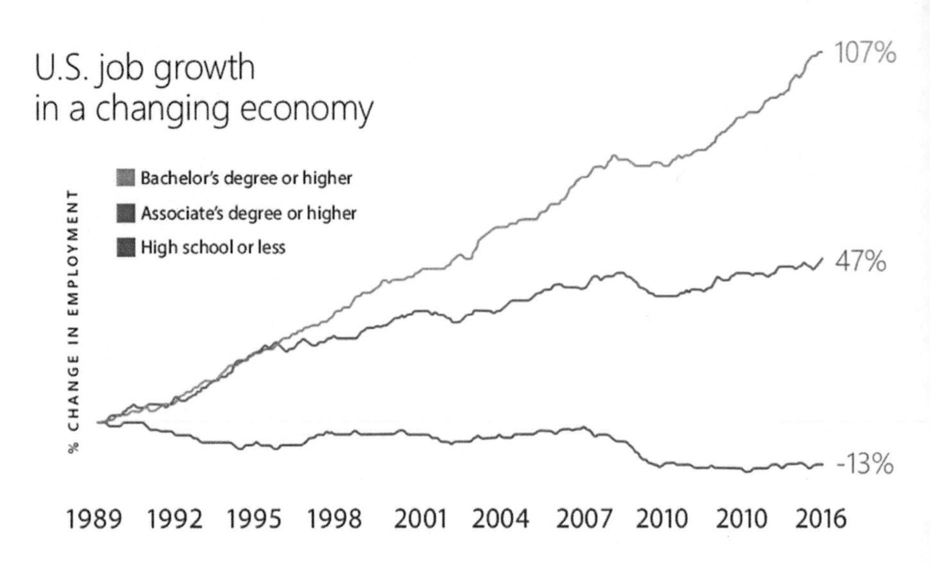

Chart 6.
Source: Georgetown Center on Education and the Workforce

technology jobs will go unfilled due to talent shortages,[30] and this gap is likely to widen given the time it takes to introduce training programs for new technology skills. According to the World Economic Forum, many academic fields experience unprecedented rates of change in core curriculum. They suggest that nearly 50 percent of subject knowledge acquired during the first year of a four-year technical degree will become outdated before students graduate. And by 2020, more than a third of the skills needed for most occupations will be ones that are not considered crucial today.[31] More broadly, technology will significantly impact the skills requirements in all job families. To manage these trends successfully, we'll need to ensure that people in the workforce can continually learn and gain new skills.

Economists who are studying the emerging talent shortage and the replacement of so-called "middle skills" jobs by automation worry that technological advances such as AI are widening the income gap between those with technological skills and training and those without.[32] As expertise in areas such as data analytics becomes more central to many jobs and automation enables machines to handle more repetitive tasks, demand for highly skilled workers will grow, and the need for those with fewer skills will fall — an effect known as the "skill-biased technical change." For example, while the number of jobs for Americans with a four-year college degree doubled between 1989 and 2016, the job opportunities for those with a high school diploma or less fell by 13 percent. Over the same period, the number of Americans with a college degree grew by just under 50 percent and the unemployment rate for those without a college degree rose 300 percent compared to those with a college degree.[33] Addressing this widening gap will require a shift in how we think about education and training so that we can prepare more of the workforce to take advantage of the opportunities that are emerging.

The Changing Nature of Work, the Workplace and Jobs

Until recently, most people worked in traditional employer-employee relationships at specific worksites: offices, factories, schools, hospitals or other business facilities. This traditional model is being upended as more people are engaged through remote and part-time work, such as contractors, or through project-based employment.

Some studies have noted that between 2005 and 2015, the number of people in alternative work relationships — which include contractors and on-demand workers — increased from 10 percent to 16 percent accounting for nearly all net job growth during that period.[34] A recent study by the McKinsey Global Institute concluded that "the independent workforce is larger than previously recognized" with up to 162 million people in Europe and the United States — 20 or 30 percent of the working-age population — engaged in some form of independent work. For more than half of these individuals, independent work supplements their primary source of income.

These alternative work arrangements are fueled by advances in technology. Perhaps the most notable trend in this regard is the rise of the on-demand economy. At its core, the on-demand economy refers to working arrangements in which people find work through online talent platforms or staffing agencies, performing tasks for a wide variety of customers. According to the McKinsey Global Institute, 15 percent of independent workers use digital talent platforms to connect to work. Researchers at Oxford University's Martin Programme on Technology and Employment estimate that nearly 30 percent of jobs in the United States could be organized into task-based work within 20 years.[35]

The on-demand economy presents enormous opportunities for workers and businesses. McKinsey estimates digital platforms that match workers with opportunities could raise global GDP by as much as 2 percent by 2025, increasing employment worldwide by 72 million full-time equivalent jobs. Here is just a partial list of the potential benefits of the on-demand economy:

- Engagement in on-demand work through digital platforms allows jobs to come to workers, rather than forcing people to migrate to available work. This helps workers who live in areas where job opportunities are limited and enables companies to access a wider talent pool.
- According to the Hamilton Project, more than 70 percent of labor force non-participants report that caregiving, disability or early retirement keeps them out of the workforce. The flexibility of on-demand work reduces the barriers that traditional employment models present.[36] According to a survey by the Pew Research Center, nearly 50 percent of on-demand workers report a "need to control their own schedule." Another quarter said there was a "lack of other jobs where they live."[37]
- The on-demand economy offers more opportunities for part-time labor. Today, many workers prefer the flexibility of part-time work to full-time employment.[38] For millennials, flexibility, work/life balance, and the social impact of their work can be more important than a high salary or a full-time career. And many baby boomers are choosing to work later in life, often through part-time work.
- The on-demand economy allows businesses to engage workers on a short-term basis, facilitating business agility and reducing long-term staffing costs. The on-

demand economy can be particularly helpful to small businesses that cannot afford a large full-time workforce but can get work done through targeted on-demand engagements. Costs can be reduced further by recruiting freelancers through online platforms that feature competitive bids for projects.

- The on-demand economy can provide companies with access to skills they do not have in-house. Hiring freelancers enables employers to find individuals with specific skills and engage them on an as-needed basis.
- The on-demand economy provides access to supplemental income. For instance, the online platform Teachers Pay Teachers includes an online marketplace where teachers buy and sell lesson plans and other educational resources.[39]

While the on-demand economy has the potential to promote greater labor force participation, many concerns have been raised about its impact on working conditions and worker protections. Some of these concerns include:

- Because the on-demand economy is so new, it is stretching the bounds of existing regulations relating to worker protections, including child labor laws and minimum wage requirements. While some on-demand digital platforms offer worker protections, others have taken the position that even baseline worker protections do not apply to the on-demand labor model.
- The on-demand borderless workplace heightens issues relating to wages and the distribution of the global workforce. Because of the differences in the

cost of living across the globe and the opportunity for employers to hire workers where wages are low, jobs may move from the higher-wage to the lower-wage countries.

- Some studies have shown that the economic benefits of the on-demand economy largely accrue to platform owners and consumers, but not to workers.[40] Because these platforms commoditize work into tasks, they may devalue other contributions that workers can make to the platform or the overall digital economy.
- The commoditization of the workforce also has the potential to reduce access to social insurance, career development and social interaction, which might otherwise strengthen innovation and economic value. Moreover, workers in the on-demand economy do not benefit from the investments enterprises make in work culture.
- In the long term, as platforms "learn" from workers and automate more tasks, the development of the platform economy may contribute to the elimination of jobs. Those who are unable to acquire new skills may be marginalized, further concentrating wealth in the hands of platform owners and top earners.

As the on-demand economy continues to grow, enterprises have an opportunity to shape policy within their own companies, at the industry level and from a public policy perspective. Increasingly, the technology industry needs to engage to change the perception that it reaps the benefits of technology progress at the expense of workers who are displaced or left without protections, benefits or long-term career paths.

Companies must acknowledge the impact of the on-demand model on workers rather than claim that they are "just the technology platform." Companies that do not acknowledge the importance of worker protections and benefits risk damage to their brands and face the possibility that lawmakers and the courts will step in to impose regulations that could limit the business opportunities that the on-demand economy presents. Microsoft believes that companies can benefit from the on-demand economy while taking steps to provide protections, benefits and opportunities that offer long-term economic stability for workers.

The technologies underpinning the on-demand economy are also changing how enterprises organize work within their traditional workforce. Today, a wide range of factors are driving enterprises to focus on creating a globally distributed workforce, including the need to look beyond local talent pools to find people with the skills that they need. But, as countries face nationalist pressures and businesses face more restrictive immigration laws, companies may also need to consider expanding their domestic workforce.

New technologies and tools are enabling businesses to accommodate distributed workforces. Online platforms can aggregate data on workers and job openings across entire countries and regions, making it easier to address geographic mismatches between skills and jobs. And because new collaboration tools support remote work, employees are no longer tied to working in a fixed location. In addition, people are seeking more flexibility in how and where they work. In a recent poll, 37 percent of technology professionals said they would take a 10 percent pay cut to work from home.[41]

While the new technologies are allowing businesses to distribute work across the globe, they require shifts in the way enterprises train workers, cultivate culture, and build institutional knowledge and intellectual property. Today, many enterprises are finding that more dispersed workforces make effective collaboration more difficult and agility more challenging. As the unit of work shifts to task-based projects that use new agile team structures, the combination of alternative employment arrangements and distributed workers means enterprises need to reconsider how they engage employees, build teams, and support career development and training. Enterprises will need to take advantage of collaboration tools like Microsoft Teams or Slack to address these shifts. They will need to use learning platforms like LinkedIn Learning or Coursera to address employees' needs for career development and mentorship. In addition, they will need to discover news ways to build community and engagement within a dispersed workforce.

Preparing Everyone for the Future of Work

Because the skills required for jobs in the AI economy are changing so rapidly, we need to ensure that our systems for preparing, educating, training, and retraining the current and future workforce also evolve. Not only will the new AI economy require new technical skills, but there is a growing recognition that most workers will need to learn new skills throughout their working lives.

According to a recent study by the Pew Research Center, 87 percent of U.S. adults in the labor force say that to keep up with changes in the workplace, it will be essential or important to get training and develop new skills throughout their working lives.[42] The ability to learn new things, collaborate, communicate and adapt to changing environments may become the most important skills for long-term employability. Innovation and new solutions throughout our education, training and workforce systems will be required to help people stay competitive in this rapidly changing workforce.

As automation and AI take on tasks that require thinking and judgement, it will become increasingly important to train people — perhaps through a renewed focus on the humanities — to develop their critical thinking, creativity, empathy, and reasoning.

Employers have a responsibility to help the education and workforce systems better understand, interpret and anticipate what professional skills they'll need. While we can't predict with certainty which jobs will exist in the future, we believe strongly that education and training will be more important than ever. Technology can be better utilized throughout the system to help students and job seekers discover promising career paths, assess their current skills, develop new skills and connect to jobs, and to scale the solutions to meet the needs of broader swaths of the population.

For people to succeed in the age of automation and AI, improving education and training systems for everyone will be critical. Most experts agree that some post-secondary education and training will be essential. The following charts show the clear relationship between educational attainment and employment levels. Chart 7 reflects this strong positive relationship in OECD countries. Chart 8 shows the United States unemployment rate impacts those with less education disproportionately and more acutely than those who accrued more education.

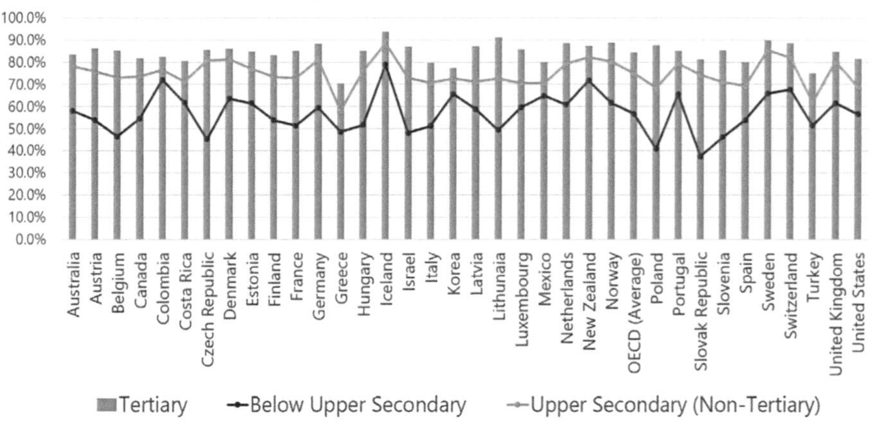

Chart 7.
Source: OECD.

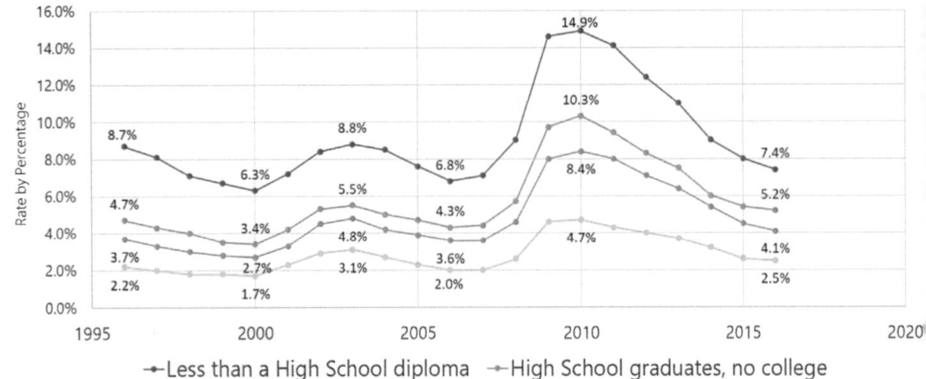

Chart 8.
Source: U.S. Bureau of Labor Statistics

The stark differences in the increases in unemployment rate, particularly for those with less education, demonstrate a higher volatility to that group. This is yet another example of how technology companies can play a vital role in shaping education and labor policy.

To help people get the training they need to thrive in today's economy and prepare for the future, Microsoft is focusing on three areas: 1) preparing today's students for tomorrow's jobs; 2) helping today's workers prepare for the changing economy; and 3) creating systems to better match workers to job opportunities.

Preparing today's students for tomorrow's jobs

The single most important skill that people will need for tomorrow's jobs is the ability to continually learn. Future jobs will require what Stanford professor Carol Dweck has called a "growth mindset" to engage in more complex problem-solving. Success will require strong communication, teamwork and presentation skills. People will need to be more globally aware as jobs will increasingly involve serving not just a community, but the world. Rapidly evolving technology impacting every sector means jobs of the future will require more digital skills, from basic computer literacy to advanced computer science. And emerging technologies and the jobs of the future will require more digital and computer skills.

Given these changing expectations, the skills young people need to learn before entering the workforce have also changed. Every young person needs to understand how computers work, how to navigate the internet, how to use productivity tools, and how to keep their computers secure. But they also need the opportunity to study computer science. Computer science teaches computational thinking, a different way to problem solve and a skill in high demand by employers. Together these skills enable access to higher paying jobs in faster-growing fields. Therefore, equitable access to rigorous and engaging computer science courses must be a top priority. If equitable access is left unaddressed, we will exclude entire populations from fully participating in this new world of work. The goal of equitable access should be computer science classrooms that are diverse across race, gender, disability and socioeconomic status.

Some countries, such as the United Kingdom, embed instruction in computational thinking into classes at every grade level, while others struggle to close the digital skills and computer science education gap. For example, while the United States has made progress to ensure that all students can take at least one computer science class before graduating from high school, thousands of students still do not have access.[43] According to the College Board, last year only 4,810 of the 37,000 high schools in the United States offered the Advanced Placement computer science exam. with girls, minorities, and the economically disadvantaged least likely to have access.[44]

To help address the global need for digital skills development, Microsoft Philanthropies partners with governments, educators, nonprofits, and businesses and is involved in a range of programs and partnerships aimed at addressing the skills gap at scale. Together with our partners, we're working to help prepare young people for the future, especially those who might not otherwise have access to opportunities to acquire critical skills. For example, through our YouthSpark program, Microsoft works with 150 nonprofit organizations in 60 countries to offer computer science learning both in and out of school to more than 3 million young people, 83 percent of whom are from underserved communities and more than half are female.

To solve this problem, increasing the number of teachers who are trained to teach computer science is also critical.

Scan for more on
TEALS

Technology Education and Literacy in Schools (TEALS) is a program that operates in 349 high schools in 29 states throughout the United States and is supported by Microsoft Philanthropies. The program engages 1,000 tech volunteers from over 500 different companies to team-teach computer science, usually with the math or science teacher. Within two years of working with their volunteer, 97 percent of classroom teachers are able to teach computer science on their own, creating the basis for sustainable computer science programs.

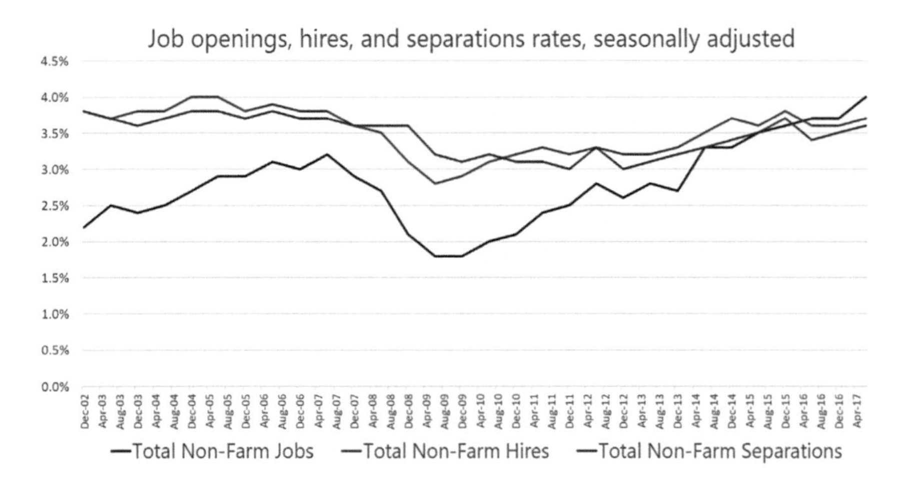

Chart 9.
Source: U.S. Bureau of Labor Statistics, Job Openings and Labor Turnover Survey, October 2017.

Supporting today's workers for the changing economy

Because technology is changing so rapidly, it's not enough to just focus on educating tomorrow's workforce; we must also help today's workers gain skills that are relevant in the changing workplace. Economic growth depends on a skilled workforce that can enable enterprises to take advantage of a new generation of emerging digital technology innovations. To achieve this, workers will need to be lifelong learners. As noted earlier, the global economy is going through rapid changes as automation and AI create demand for a more skilled workforce. This is reflected in recent labor statistics in the United States where, for the first time, job postings have surpassed hiring in the monthly U.S. Bureau of Labor Statistics Job Openings and Labor Turnover Survey (JOLTS) reporting.[45]

This is just one illustration of the global mismatch between employer needs and the skills that today's workers possess. According to a 2017 survey by global staffing firm ManpowerGroup, significant skills shortages exist in Japan, India, Brazil, Turkey, Mexico, Greece, Australia and Germany.[46] In the United States, the National Skills Coalition reports that 53 percent of jobs today are "middle skill" or "new-collar" jobs that require more than a high school diploma and less than a college degree. But only 43 percent of the workforce is a match for this requirement. At the same time, while 20 percent of the workforce has a high school graduation credential or less and is considered "low-skilled," just 15 percent of jobs are open to people with this level of educational attainment.[47] Further, in a study of job postings by Burning Glass Technologies, 8 out of 10 middle-skills jobs require basic digital literacy skills, something that more than half of workers today lack. Unless we change how we prepare people for these new jobs, this gap will continue to widen.[48] The National Skills Coalition predicts that 80 percent of jobs that will be created by 2024 will require post-secondary credentials.[49]

As demands for a more educated and skilled workforce continue to grow, we must identify new ways to increase the skills of today's workers. Workforce systems will need to evolve to keep pace with the changing technologies. Emerging practices focused on distance and online learning as well as investment in more on-the-job training programs are key ways to prepare today's workers for the changing workplace.

To understand how to train the current workforce, it is important to identify the skills that enterprises need. Microsoft and its LinkedIn subsidiary are already experimenting with new ways to understand which skills are currently in demand and how to help people gain them.[50] For example, LinkedIn is working with the National Cybersecurity Center (NCC) and the University of Colorado at Colorado Springs to identify the most in-demand cybersecurity occupations in the United States and map the skills needed to be hired for those jobs.

LinkedIn is also working with local training programs to update curriculum and to teach graduating students how to use LinkedIn in their job search. Microsoft offers curriculum and certification programs to help people develop digital skills through programs like Imagine Academy, YouthSpark and LinkedIn Learning.[51] This is important because digital skills are critical in all job clusters. In fact, research firm IDC reports that knowing how to use Microsoft Office was the third most cited skill requirement across all occupations.[52]

It will also be critical to identify new entry points into the workforce. As enterprises face talent shortages, they must explore new ways to bring in talent from available labor pools. Microsoft and LinkedIn are testing several programs that do this, such as Microsoft Software and Systems Academy (MSSA). An 18-week training program, MSSA is created specifically to prepare service members and veterans for careers in cloud development, cloud administration, cybersecurity administration, and database and business intelligence administration. At the end of the program, graduates interview for full-time jobs at Microsoft or one of our hiring partners. So far, 240 companies have hired graduates of MSSA. Microsoft is also working with the state of Washington's Apprenticeship and Training Council, which offers the first registered apprenticeship program for the IT industry.

Scan for more on
MSSA

LinkedIn supports the apprenticeship model as well, and is working to identify ways to build an apprenticeship marketplace. LinkedIn launched REACH, a six-month apprenticeship program where people join a LinkedIn engineering team to learn what it is like to work as a software engineer and gain experience to help them pursue a career in software development. LinkedIn is also partnering with CareerWise Colorado to create a marketplace that lists apprenticeship openings for high school students. And it is working with the state of Colorado's apprenticeship office to help people understand the value of apprenticeships.

All of these programs are good first steps. But the next — and maybe harder — challenge will be to figure out how to scale these programs through public and private partnerships to have sustainable impact on the workforce. This will require both educational institutions to think differently — at scale — about how they train, and employers to think differently about how they identify and onboard talent.

Supporting the development of systems to create a skills-based marketplace

To help foster economic prosperity across the globe, the public and private sectors must also invest in new educational delivery models. People need to be able to gain the demand-driven skills required for advancement and a system needs to be developed in which credentials are portable, stackable and valued by employers. The rapidity of change in the workplace requires employers and workforce providers to work together in new ways. The public and private sectors should seek to meet the needs of people at all stages of the workforce continuum — from students entering the workforce to unemployed and underemployed workers, to people currently in the workforce who need help gaining new skills to ensure their long-term employability.

To help enterprises find qualified employees and workers find jobs, we'll need to shift from a system based on traditional degrees to a system based on skills. This system should account for the rapidly evolving skills employers need across occupations. And it should recognize the skills individuals possess to more efficiently connect workers to employers.

A first step will be to create a common taxonomy of skills. Emerging technologies and changes to the workplace require education providers to offer training in the skills that employers need. It will be critical to codify the most in-demand skills and train workers on them and on how to articulate their skills to potential employers.

Employers and workforce agencies should use real-time labor market information to identify in-demand skills, a task for which LinkedIn and the broader IT industry are well-placed to assist governments and workforce agencies. Governments can use this information to develop and deliver high-quality workforce training programs and offer incentives and financial resources to private and nonprofit organizations to provide training. Goals for educational attainment should include outcomes related to employment, skills and advancement.

Technology and data must be used to build a dynamic skills-based labor marketplace that guides the education and workforce systems. To be successful, we'll need a worker-centered framework for assessing learning outcomes that harmonizes data across sectors in a way that is easier for individuals to navigate. This framework should emphasize the knowledge that employers require and include the technical and foundational skills workers need in the digital workplace. Foundational skills include problem-solving, work ethic, teamwork, curiosity and interpersonal communication. This framework should guide training organizations as they help people acquire skills and earn credentials.

We also need to identify existing open jobs and the skills required to fill them. Digital platforms such as LinkedIn, TaskRabbit and Upwork offer insights about in-demand skills based on job or task openings. Over time, this data can be used to construct analyses such as the LinkedIn Economic Graph to understand supply and demand for specific skills and how they vary over time for a given region, particularly when combined with government data on local demographics and businesses.

Microsoft and LinkedIn are taking additional steps to understand which skills are in high demand, to invest in skills development to address the changing nature of work and of jobs, and to help people find jobs to match their skills. In pursuit of these goals, Microsoft and LinkedIn have partnered with Skillful, an initiative of the Markle Foundation, that is creating a skills-based labor market that works for everyone, with a focus on those without a college degree. Microsoft has made a substantial investment to help the Markle Foundation build this marketplace.[53]

To achieve its mission, Skillful helps employers expand their talent pool by providing data, tools and resources to simplify the adoption of skills-based practices. Coaches and digital services enable job seekers to find out which skills are in demand and access professional training at any stage of their careers. Skillful also works with educators and employers to ensure that students are learning the skills they need to succeed in today's digital economy. The partnership aims to create a model that can be replicated across the United States, aiming to help millions of Americans find rewarding careers. Skillful is also working with LinkedIn to test strategies to improve the lives of skilled workers through initiatives such as Mentor Connect, LinkedIn's pilot mentorship program that uses Skillful's coaching efforts and platform.

To improve how the public and private sectors work together to match job seekers with job openings, LinkedIn has opened its listings to governments in the United States, free of charge. In 2017, more than 1 million government job listings appeared on LinkedIn. In addition, the National Labor Exchange, which is managed by the National Association of State Workforce Agencies and includes jobs from all 50 state job banks, began sending its jobs to LinkedIn in January 2017. LinkedIn has shared labor market insights with more than 70 U.S. cities through the White House TechHire program. LinkedIn has also shared data with government agencies in New York, Los Angeles, Chicago, Louisville, New Orleans, Seattle, San Francisco and Cleveland to help them improve issues such as student retention and youth unemployment, identify job biases, and understand supply and demand for job skills.

While many of these programs are relatively new, it is clear that we need to use data to build a more dynamic skills-based labor marketplace that guides education and workforce systems and prepares workers for available jobs.

Changing Norms for Changing Worker Needs

To meet the challenges of the evolving economy, we must also understand how the on-demand economy, part-time work, independent contracting, and temporary jobs affect individuals and society.[54] These changes raise questions that are not always adequately addressed by existing legal and policy frameworks.

To enable innovation and to protect workers, the public and private sectors must tackle a number of key policy questions. Legal certainty must be created so that workers and businesses understand their rights and obligations. Industry must also define its own standards for worker protection to ensure that society does not become further divided between the "haves and have-nots." To promote the efficient flow of skills, encourage entrepreneurship, and allow workers to exercise their market power to the best of their ability, industry and governments must work together to find ways to enable workers to take their benefits with them as they change jobs. And the social safety net must be modernized to support workers and families, as well as stabilizing the economy during periods of economic instability and labor market shifts.

Providing Legal Certainty and Structure for Employers and Workers

Given the velocity of change in the modern workforce, it is not surprising that existing legal and policy frameworks do not adequately address all of today's changing work arrangements. Questions and uncertainty about how to categorize workers have been an issue for a sometime, with consequences for businesses, workers and government. Now, changes in labor marketplaces and the rise of on-demand work platforms are increasing the urgency to find answers to these questions.[55]

Broadly speaking, current laws tend to recognize only two designations for workers: 1) employees who work on a regular basis in a formal relationship with an employer; or 2) independent contractors who provide goods or services under a specified contract.

Employees have traditionally enjoyed less flexibility and control over their hours and working conditions, but retain more stability and legal protection. Independent contractors typically retain more control over when and how they work, but receive fewer legal protections. Whether or not someone is an employee determines whether they are protected by traditional labor, wage and hour, and equal opportunity laws, and whether they can access employer-provided benefits such as private pensions, access to training, retirement benefits and, in many countries, healthcare. A worker's designation also determines whether employers contribute to and workers benefit from social safety net benefits such as unemployment insurance and, in the United States, Social Security and state-paid leave benefits.

Today, most on-demand workers are treated as independent contractors by digital platforms and the businesses that engage them. Under this classification, on-demand workers are not protected by minimum wage and overtime pay requirements, child labor regulations, or anti-discrimination and anti-harassment laws. In addition, there is a lack of clarity about the rights and protections that workers who connect through an intermediary can expect under the law. As dissatisfaction about the lack of protections grows, on-demand workers are increasingly challenging such designations through litigation or government intervention.[56] The results have been inconsistent. For on-demand workers, this creates uncertainty about what rights and benefits they can expect. For platform companies and the businesses that engage on-demand workers, it raises questions about whether on-demand workers will be considered to be employees, subject to the associated costs and protections.

Until labor and employment laws and systems for providing benefits are modernized to respond to current workforce trends, there's a danger that growth in productivity and opportunity will be constrained. There is a risk that if we fail to impose baseline protections — including wage protection — work will become increasingly stratified between high paying, stable employment and low-value, low-paid, task-oriented gigs. This may undermine the potential of the on-demand economy. Unfortunately, current discussions about the classification of workers are often extremely polarized — with business pushing for narrower classifications and labor advocates pushing for more expansive interpretations. What is needed is broader dialogue about the needs of businesses and workers to determine what changes are required to serve the interests of both in a way that is productive and fair.

So far, policy recommendations have focused on either redefining the categories of employees and independent contractor or finding ways to mitigate the consequences of this difference — often by extending protections, benefits and social safety net participation to contingent workers. Both of these approaches focus on addressing the issue by making the distinction between the two categories less extreme and providing basic protections to workers who are currently left out. Current policy proposals include the implementation of a new worker classification for "independent worker" that would fall between employee and independent contractor; the creation of a safe harbor for income and employment tax purposes for certain workers; the expansion of collective bargaining and other protections to certain classifications of on-demand workers; and the adoption of voluntary minimum industry standards for worker protections. All of these proposals should be explored as more people find work through on-demand platforms.

Developing Industry Standards to Protect All Workers

Today, business leaders have the opportunity to play a significant role in reshaping employment policy for the emerging economy by setting their own standards for on-demand engagements. Microsoft believes we can (and should) positively impact the treatment of on-demand workers through its internal policy. Microsoft's policy includes minimum pay requirements for all on-demand work. It requires that on-demand workers be paid within one week of completing work and that all workers be treated with dignity and respect. It also prohibits the use of child labor and requires the on-demand platforms that it uses to be accessible. Microsoft is implementing contractual terms with the on-demand platforms it engages with that reflect this policy.

While corporate policies can provide some degree of protection to on-demand workers, the impact is limited. But enterprise users of on-demand labor also have an opportunity to contribute to broader solutions to these issues. For instance, groups such as freelancers' unions and caregiver coalitions have improved standards for task workers — sometimes through legislation. Approaches like the National Domestic Workers Alliance's "Good Work Code" for domestic workers in the United States offer a framework for engaging workers that includes safety, shared prosperity, a living wage, inclusion and input.[57] Industry leaders should encourage discussions among businesses and workers to develop standards like these for task-based work that might include wage, benefits and fair treatment commitments. This could lead to a set of standards endorsed by businesses that might serve as a framework for nongovernmental policy. Such standards could either be industry-specific or generalized to broader platforms, and might also serve as the framework for legislation that sets minimum protections.

Ensuring Benefits Move with Workers

These labor market trends have tremendous implications for both worker protections and employer-provided benefits. The employer-based benefits model that emerged in most of Europe and North America in the middle of the last century is based on two principles: first, that businesses benefit from the well-being of a stable workforce; and second, that certain benefits are best provided by employers rather than the government as an investment in workforce stability.

This approach has shaped our perspective of the social contract between employers and employees. While the nature of work has evolved with technology innovation, the system of employer-provided benefits and social safety nets has not. The challenge we face now is how to transform benefits and social insurance programs to provide adequate coverage for workers and a sustainable contribution structure for businesses.

In today's digital economy, the mobility of labor and the ability to quickly focus skills on new growth areas are vitally important to business success. Many businesses may find the relative burden of maintaining employer-provided benefits not worth the cost. Individual workers also want benefits that are portable and flexible. Portability of benefits will be critical to a viable solution. Three models have emerged as possible solutions.

- **Employer-provided benefits.** The issue of providing benefits to people working in industries that are structured around short-term projects is not new.

Industries such as construction and entertainment have addressed this through labor-management partnerships that enable workers to retain healthcare and pension coverage across multiple employers, even for short-term work. A collective bargaining structure provided a way for employers to contribute to benefits pools without bearing the burden of administration; workers did not have to be responsible for moving benefits and seeking out new providers. New models could use this approach, which would reduce inefficiency and confusion, and ensure that workers have access to basic protections and adequate benefits. This would support greater labor mobility because workers would be less likely to stay in jobs simply to retain benefits.

- **Use of new platforms to provide benefits.** The rise of on-demand labor platforms may create opportunities to develop new ways for workers to access benefits. For example, Care.com, a platform for caregivers, enables families to contribute to their caregiver's benefits in a way that is similar to how traditional corporate employers fund employee benefits.[58] When families pay a caregiver through Care.com, a percentage funds benefits that stay with caregivers even when they go to work for other families on Care.com. There are still challenges to this approach — including what happens when workers find work through different platforms.
- **Government mandates and funds.** In some countries, national or even multinational government organizations may seek to address this gap. In those countries where a broad new nationwide system may not be feasible, smaller governmental units may be able to

establish the infrastructure and risk pooling needed to make benefits affordable. Some countries require basic benefits, with an accompanying structure to provide those benefits. In the United States, where broad new federal programs have not received political support, some states have sought to create their own healthcare or retirement programs. In the short term, policymakers should consider creating pilot programs to establish portable benefits, such as, legislation introduced at the state level in the United States.[59]

Modernizing the Social Safety Net

A more mobile and dynamic workforce will increase pressure on social safety net programs. As people find work through a more diverse array of non-exclusive arrangements that may not include employer-provided benefits or allow workers to earn enough to build their own savings, they will rely more than ever on safety net programs like unemployment insurance, workers' compensation and Social Security.

Programs that are triggered during a worker's productive working years are particularly important for workers' economic stability, which in turn helps maintain a diverse and skilled workforce. Periods of joblessness produce income volatility, which can have serious long-term consequences for workers and their families. This also reduces the pool of available skilled labor for businesses. Even a robust economy includes a significant level of under-employment or unemployment. In August 2017, the U.S. Bureau of Labor Statistics estimated that 7.1 million American workers were unemployed, with an additional 5.3 million working part-time for economic reasons or as involuntary part-time workers. These periods are likely to occur many times over a worker's life.[60]

Many existing social safety net programs are already underfunded and face further fiscal pressures as workforces age. This means that during periods when there is increased need, such as during a recession, existing safety nets are likely to prove inadequate. Compounding the problem, many comprehensive safety net programs are heavily dependent on traditional employment relationships. A significant shift away from traditional employment without corresponding policy changes could further erode work-based social safety net programs. Finally, these programs do not take into account newer models of work, nor do they anticipate that individuals may move in and out of the workforce with greater frequency or for a greater variety of reasons. It will be essential to modernize these programs to encourage labor mobility and enable workers to gain new skills and connect to new opportunities.

Companies can begin to experiment with public-private partnerships to explore how to meet the needs of workers. For example, Microsoft, through LinkedIn, is exploring new ways to speed the re-employment of workers in the United States. LinkedIn is working with the state of Utah to test network-based job searching as a strategy for reemployment through a pilot program that was recently highlighted by the Trump administration for saving taxpayer money by enabling unemployed workers to find new jobs more effectively.
In addition, Microsoft and LinkedIn are building tools for employment counselors and job seekers that would improve

workforce programs such as unemployment insurance and state workforce programs. And LinkedIn is working with the National Association of State Workforce Agencies to produce job search curriculum for its network of 2,500 publicly managed job centers in the United States.

Enterprises should continue to use data and technology tools to assist governments in identifying opportunities for worker redeployment to scale these solutions beyond pilots and experiments. However, modernizing the social safety net will require a multifaceted approach such as:

- **Rethink unemployment insurance and reemployment programs, including job training and trade adjustment assistance programs.** Steps have been proposed to begin modernizing unemployment insurance and to bolster the program's solvency. Businesses should engage in discussions about the importance of next-generation versions of unemployment insurance and employment services that take into account newer models of work; anticipate that individuals may move in and out of the workforce with greater frequency; promote greater labor mobility; and help workers gain new skills and connect with new opportunities.
- **Reform tax policy and social safety net.** Policymakers must explore how to adjust policies to adequately fund social safety net programs. This may include going beyond existing tax bases to consider other methods of funding social safety nets. For example, some have questioned whether wages are the right measure of income to be taxed.

> Where business productivity may be better measured by production than through wages, some propose assessing taxes to support social safety nets and government revenue based on other measures.

The case must also be made for how social programs can increase the size of the labor pool; be structured to help employees move in and out of work more easily and more flexibly; and reduce burdens for employers. Without significant modernization, social safety nets will not adequately support emerging models of work. The private and public sectors must join together to explore how to best support workers in the new economy.

Working Together

As we move forward, it will be essential for governments, the private sector, academia, and the social sector to join together to explore how to best support workers in the new economy. This can be achieved by developing new approaches to training and education that enable people to acquire the skills that employers need as technology advances; by creating innovative ways to connect workers with job opportunities; and by modernizing protections for workers to promote labor mobility and cushion workers and their families against uncertainty in a fast-changing global economy.

Conclusion

AI Amplifying Human Ingenuity

What happens when we begin to augment human intelligence and ingenuity with the computational intelligence of computers? What does human-centered AI look like?

It may look a lot like Melisha Ghimere, a 20-year-old computer science student at Kantipur Engineering College in Kathmandu, Nepal. Melisha's team was a regional finalist in Microsoft's Imagine Cup competition in 2016.

Like the vast majority of the people of Nepal, she comes from a family of subsistence farmers who raise cows, goats and water buffalo. Over the years, her aunt and uncle, Sharadha and Rajesh, did well, building a herd of more than 40 animals — enough to raise two children, support four other relatives, and even hire a few workers to help out. But then, seven years ago, an outbreak of anthrax wiped out much of their herd. They are still struggling to regain their economic footing.

At college, Melisha's family was never far from her mind. So she set out to develop a technology-based solution that would help farmers like her uncle. Working with three other students, she researched livestock farming and veterinary practices, and spoke with many farmers. Together, they built a prototype for a monitoring device that tracks temperature, sleep patterns, stress levels, motion and the activity of farm animals. Melisha's AI system predicts the likely health of each animal based on often subtle changes in these observations. Farmers can follow the health of their animals on their mobile phones, access advice and recommendations to keep the animals healthy, and receive alerts when there are signs of sickness or stress, or when an animal might be pregnant.

Melisha's project is still in its infancy, but the early results have been promising. In the first field tests, the solution was about 95 percent accurate in predicting an animal's health. It already enabled one family to prevent a deadly outbreak by identifying a cow that was in the earliest stages of an anthrax infection, before symptoms were evident to the farmer.

Like Melisha's project, AI itself is still at a nascent stage. Thanks to advances in the past few years, we're beginning to build systems that can perceive, learn and reason, and on this basis, can make predictions or recommendations. Nearly every field of human endeavor could benefit from AI systems designed to complement human intelligence. From preventing once-deadly diseases, to enabling people with disabilities to participate more fully in society, to creating more sustainable ways to use the earth's scarce resources, AI promises a better future for all.

Change of this magnitude inevitably gives rise to societal issues. The computer era has required us to grapple with important questions about privacy, safety, security, fairness, inclusion, and the importance and value of human labor. All of these questions will take on particular importance as AI systems become more useful and are more widely deployed. To ensure that AI can deliver on its promise, we must find answers that embrace the full range of human hopes, needs, expectations and desires.

This will take a human-centered approach to AI that reflects timeless values. And it will take an approach that is firmly centered on harnessing the power of computational intelligence to help people. The idea isn't to replace people with machines, but to supplement human capabilities with the unmatched ability of AI to analyze huge amounts of data and find patterns that would otherwise be impossible to detect.

How AI will change our lives — and the lives of our children — is impossible to predict. But we can look to Melisha's device — a device that could help millions of small farmers in remote communities live more prosperously — to see one example of what can happen when human intelligence and imagination are augmented by the power of AI.

We believe there are millions of Melishas around the world — people young and old who have imaginative ideas for how to harness AI to address societal challenges. Imagine the insight that will be unleashed if we can give them all access to the tools and capabilities that AI offers. Imagine the problems they will solve and the innovations they will create.

This won't happen by itself. A human-centered approach can only be realized if researchers, policymakers, and leaders from government, business and civil society come together to develop a shared ethical framework for artificial intelligence. This in turn will help foster responsible development of AI systems that will engender trust. As we move forward, we look forward to working with people in all walks of life and every sector to develop and share best practices for building a foundation for human-centered AI that is trusted by all.

Endnotes

1. See Harry Shum blog, July 2017 at https://blogs.microsoft.com/blog/2017/07/12/microsofts-role-intersection-ai-people-society.

2. https://blogs.microsoft.com/ai/microsoft-researchers-win-imagenet-computer-vision-challenge.

3. https://www.microsoft.com/en-us/research/blog/microsoft-researchers-achieve-new-conversational-speech-recognition-milestone.

4. See Harry Shum blog, May, 2017 at https://blogs.microsoft.com/blog/2017/05/10/microsoft-build-2017-microsoft-ai-amplify-human-ingenuity.

5. https://www.microsoft.com/en-us/research/project/medical-image-analysis.

6. https://www.microsoft.com/en-us/research/project/project-premonition.

7. For example, when you ask Cortana "How big is Ireland?" the response is not only in square kilometers, but also says "about equal to the size of South Carolina."

8. https://www.microsoft.com/en-us/seeing-ai.

9. https://www.microsoft.com/en-us/research/project/farmbeats-iot-agriculture/#.

10. https://www.partnershiponai.org.

11. https://www.nytimes.com/2017/10/26/opinion/algorithm-compas-sentencing-bias.html and https://www.propublica.org/article/machine-bias-risk-assessments-in-criminal-sentencing.

12. https://www.nytimes.com/2017/11/21/magazine/can-ai-be-taught-to-explain-itself.html.

13. Daniel Solove, "A Brief History of Information Privacy Law," [GW Law] 2006, p.1-25.

14. One interesting set of insights emerges from the transition from horses to automobiles. This gave birth to multiple new industries, many of which were impossible to predict when cars first came into use. https://www.linkedin.com/pulse/today-technology-day-horse-lost-its-job-brad-smith.

15. http://www3.weforum.org/docs/WEF_FOJ_Executive_Summary_Jobs.pdf.

16. http://query.nytimes.com/gst/abstract.html?res=

9C03EEDF1F39E133A25755C2A9649C946995D6CF&legacy=true.

17. https://www.economist.com/news/special-report/21700758-will-smarter-machines-cause-mass-unemployment-automation-and-anxiety.

18. https://www.economist.com/news/special-report/21700758-will-smarter-machines-cause-mass-unemployment-automation-and-anxiety

19. https://www.economist.com/news/special-

report/21700758-will-smarter-machines-cause-mass-unemployment-automation-and-anxiety.

20. https://venturebeat.com/2017/10/04/the-fundamental-differences-between-automation-and-ai.

21. https://www.washingtonpost.com/news/theworldpost/wp/2017/10/19/inside-chinas-quest-to-become-the-global-leader-in-ai/?utm_term=.9da300d7d549.

22. AI Survey. Risk Drivers. https://news.microsoft.com/cloudforgood/policy/briefing-papers/responsible-cloud/amplifying-human-ingenuity-artificial-intelligence.html.

23. https://www.oxfordmartin.ox.ac.uk/downloads/academic/The_Future_of_Employment.pdf.

24. https://openknowledge.worldbank.org/handle/10986/23347.

25. https://papers.ssrn.com/sol3/papers.cfm?abstract_id=2940245.

26. https://www.theguardian.com/technology/2017/jan/11/robots-jobs-employees-artificial-intelligence.

27. https://www.postandcourier.com/business/as-amazon-pushes-forward-with-robots-workers-find-new-roles/article_c5777048-97ca-11e7-955e-8f628022e7cc.html.

28. https://www.forrester.com/report/The+Future+Of+Jobs+2025+Working+Side+By+Side+With+Robots/-/E-RES119861.

29. https://www.economist.com/news/special-report/21700758-will-smarter-machines-cause-mass-unemployment-automation-and-anxiety.

30. "The new new way of working series: Twelve forces that will radically change how organizations work," BCG, March 2017. https://www.bcg.com/en-us/publications/2017/people-organization-strategy-twelve-forces-radically-change-organizations-work.aspx.

31. http://reports.weforum.org/future-of-jobs-2016/skills-stability/?doing_wp_cron=1514488681.1306788921356201171875.

32. https://www.technologyreview.com/s/515926/how-technology-is-destroying-jobs.

33. https://cew.georgetown.edu/wp-content/uploads/Americas-Divided-Recovery-web.pdf.

34. https://krueger.princeton.edu/sites/default/files/akrueger/files/katz_krueger_cws_-_march_29_20165.pdf.

35. http://www.oxfordmartin.ox.ac.uk/publications/view/1314.

36. http://www.hamiltonproject.org/papers/who_is_out_of_the_labor_force.

37. http://www.pewinternet.org/2016/11/17/gig-work-online-selling-and-home-sharing.

38. According to the Bureau of Labor Statistics, 6 million people are working part-time because that is their preference, an increase of 12 percent since 2007. http://www.bloomberg.com/news/articles/2015-08-18/why-6-million-americans-would-rather-work-part-time.

39. https://www.teacherspayteachers.com.

40. http://journals.sagepub.com/eprint/3FMTvCNPJ4SkhW9tgpWP/full.

41. http://globalworkplaceanalytics.com/resources/costs-benefits.

42. http://www.pewsocialtrends.org/2016/10/06/4-skills-and-training-needed-to-compete-in-todays-economy.

43. Furthermore, according to the National Center for Education Statistics, 1 in 5 high school students does not graduate within 4 years of beginning high school.

44. https://secure-media.collegeboard.org/digitalServices/pdf/research/2016/Program-Summary-Report-2016.pdf.

45. https://www.bls.gov/charts/job-openings-and-labor-turnover/opening-hire-seps-rates.htm.

46. https://www.bloomberg.com/news/articles/2017-06-22/the-world-s-workers-have-bigger-problems-than-a-robot-apocalypse.

47. https://www.nationalskillscoalition.org/resources/publications/2017-middle-skills-fact-sheets/file/United-States-MiddleSkills.pdf.

48. http://burning-glass.com/wp-content/uploads/2015/06/Digital_Skills_Gap.pdf.

49. https://www.nationalskillscoalition.org/resources/publications/file/Opportunity-Knocks-How-expanding-the-Work-Opportunity-Tax-Credit-could-grow-businesses-help-low-skill-workers-and-close-the-skills-gap.pdf.

50. The availability of broadband in remote and underserved communities can be instrumental in expanding the quality and accessibility of educa-

tion, training and broader civic engagement. But there are 23.4 million people living in rural counties who don't have access to broadband and therefore do not have access to on-demand learning tools. To meet that need, in July 2017, Microsoft launched its Rural Airband Initiative to help serve as a catalyst for broader market adoption of this new model and to eliminate the rural broadband gap in the U.S. by July 4, 2022.
https://news.microsoft.com/rural-broadband.

51. One example of Microsoft's global skills initiatives is Microsoft India's Program Oorja, which works with polytechnics, industrial technology institutes and engineering colleges to enable students to be ready for work by helping them acquire certifications in various Microsoft Education curricula, largely in office productivity.
https://www.microsoft.com/en-in/about/citizenship/youthspark/youthsparkhub/programs/partners-in-learning.

52. https://news.microsoft.com/download/presskits/education/docs/IDC_101513.pdf.

53. https://news.microsoft.com/2017/06/27/the-markle-foundation-and-microsoft-partner-to-accelerate-a-skills-based-labor-market-for-the-digital-economy.

54. Just as more accurate and up-to-date data is needed to understand evolving jobs and needed skills, more data also is needed to better understand how employer and employee relationships and working conditions are evolving, including how the nature of work is changing. In addition, many existing government programs rely upon wage data to assess employment outcomes; a broader set of data may be needed to understand the true impact of newer contingent worker arrangements. Platform companies can contribute private-sector data to enhance this analysis.

55. Although online platforms, by most estimates, still only make up less than 1 percent of the workforce, the percentage of workers not in traditional employer/employee work arrangements (temporary agencies, on-call workers, contract workers, independent contractors or freelancers) is much greater. See, e.g., The Rise and Nature of Alternative Work Arrangements in The United States, 1995-2015.

56. In the absence of modernized laws, regulatory agencies are developing interpretations that represent vast departures from prior precedent — for example, expanding the scope of joint employment. With the changing political composition of many regulatory agencies, there is the potential for new case law that swings the pendulum in the opposite direction. The United States Congress is also proposing to legislate key definitions.

57. http://www.goodworkcode.org/about.

58. http://www.care.com.

59. See, e.g., S. 1251 and H.R.2685, Portable Benefits for Independent Workers Pilot Program Act, introduced by Senator Warner and Rep. DelBene. The act would establish a portable benefits pilot program at the U.S. Department of Labor, providing $20 million for competitive grants for states, local governments and nonprofits to pilot and evaluate new models or improve existing ones to offer portable benefits for contractors, temporary workers and self-employed workers.

60. We know from existing data that workers in recent decades already experience multiple instances of joblessness over a career. The National Longitudinal Survey of Youth 1979 (NLSY79) tracked a nationally representative sample of people born in the years 1957 to 1964; they experienced an average of 5.6 spells of unemployment from age 18 to age 48. High school dropouts experienced an average of 7.7 spells of unemployment from age 18 to age 48, while high school graduates experienced 5.4 spells and college graduates experienced 3.9 spells. In addition, nearly one-third of high school dropouts in the survey experienced 10 or more spells of unemployment, compared with 22 percent of high school graduates and 6 percent of college graduates.

Made in the USA
San Bernardino, CA
12 May 2018